THE *UFO* EXPERIENCE

RECONSIDERED

SCIENCE and SPECULATION

by

Robert L. Mason

Printed in the U.S.A.
First edition

All artwork by Robert L. Mason
(with the exception of Figure #2)

For inquiries and order information contact:
Schooner Moon Books
Robert L. Mason
P.O. Box 377
Mendocino, CA 95460
rlmason@mcn.org

ISBN 978-0-6151-9045-7

Cover Art:
Particle Beam with Entanglement by Robert L. Mason

To Deanna, Tim,
and Cora

Table of Contents

List of Illustrations and Figures

Acknowledgements

First and foremost I wish to thank Gordon Chism without whose persistence I never would have taken this subject seriously. Next, I am indebted to my reviewers. For science and technology: Paul Soderman, a retired NASA engineer, and my son-in-law Tim Becker, a geologist with the Berkeley Geochronology Center. For the nuts and bolts of the English language: Evie Wilson, my partner-in-life, and fellow artist/ writer, plus my daughter, Deanna Mason, an attorney in Berkeley, California. For help with computer graphics and various aspects of the publishing process, I have relied on Rick Droz who is our neighbor in Mendocino, California, and Kathy Carl of Avenue Design in Fort Bragg, California. Finally, I am grateful for the splendid index created by Medea Minnich of Mendocino, California.

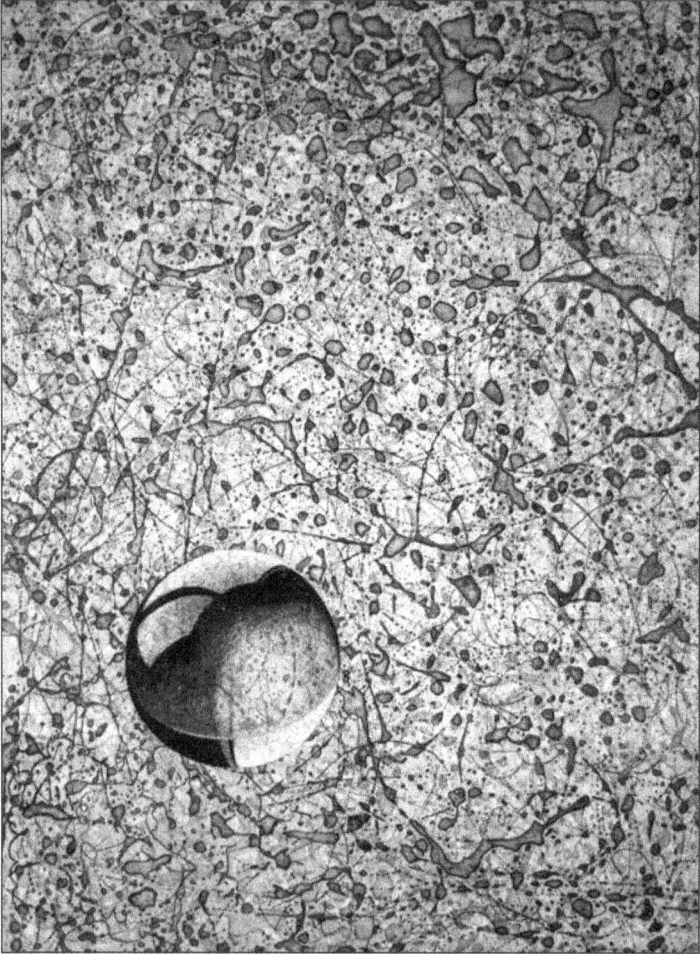

monoprint   A Signal in the Noise   R. L. Mason

Introduction

Let me admit, right from the start, that this field of study is extraordinarily polluted. The paucity of hard facts has allowed imaginations to run wild, hoaxes to be perpetrated, fraud, and outright lies. In addition, a large number of reports (perhaps the majority) are simply misidentifications or misinterpretations of familiar objects or phenomena. *But is there a signal hidden in all this noise?*

UFO's exist, of course, that is, Unidentified Flying Objects exist. If a flying object is observed which cannot be identified then it is a UFO by definition. The *key* question is whether there is an extraterrestrial intelligence behind any of them, and that is where the confusion starts. Go into any book store and ask where the shelf containing books on this subject is, and you will receive directions to the section on the occult, mysticism, the paranormal, and metaphysics. There, nestled in with Bigfoot and poltergeists you will find the books on the subject at hand. I don't see how there could be any clearer testimony to the feeling of the general public (not to mention the scientific community) when the subject of UFO's is broached. Even those who bill themselves as UFO investigators or "Ufologists" are a pretty erratic lot, tending toward sensationalism, continually quoting or disagreeing with each other. You have to move mountains of detritus to find a few flecks of gold.

Some authors on this subject have been proven to be fraudulent. Long before space probes established that the planet Venus has a surface temperature that can melt lead, and is an arid inferno of crushing atmospheric pressure and acidic clouds, George Adamski claimed, in the book *Flying Saucers Have Landed* [07], to have had a conversation with a Christ-like alien being who informed him he was visiting from Venus.[1]

All of the above has caused serious, scientific investigators to avoid the subject like the plague. I am not a scientist; I'm an engineer with some exposure to particle accelerators. Being an engineer probably gives me a slightly different slant on things. Engineers are usually more concerned with how things can be made to work (technology) as opposed to the basic nature of things (science). The reader may detect that inclination in this book.

Until I read *The UFO Experience: A Scientific Inquiry*[13] by J. Allen Hynek I was quite skeptical concerning the whole subject. *Since* reading the book I have moved from negative to neutral, or agnostic and *very* curious. Dr. Hynek convinced me that some physical phenomena is being observed that deserves serious investigation. But what is it? In *this* book speculations are offered in three general categories: natural causes, domestic technology, and alien technology. But perhaps more importantly, a new way of looking at the phenomena is proposed that finds itself at home in any of these three possibilities. More space has been given to possible alien technologies because of the far-reaching implications for the human race.

Writing this book has been a personal adventure of the mind which is why it is written in the first person. The tale unfolds

---

[1] [07] p. 198

roughly in the sequence of the adventure with recollections and research inserted where they seemed to fit best. The initial motivation grew out of ten years worth of conversation with Gordon Chism whose own adventure is included in Chapter 2. Gordon is a firm believer in an alien presence, and if I had been in his shoes, I might be too. However, I am a natural born skeptic. I am sure Gordon's account is accurate, but I had doubts about his interpretation. So I had to ask myself "Okay, what *did* he see?" Chapter 3 is a summary of my first attempt at answering that question, and although it was an interesting exercise, it ended up inconclusive. In succeeding chapters I wander off in other directions, explore other avenues, and learn much in the process. Eventually, in Chapter 13, I come full circle and return to my starting point armed with new knowledge and what I believe is a unique perspective.

In short, what we have here is a puzzle. I have always enjoyed trying to solve puzzles, and in this case the reader is invited along for the adventure.

Chapter 1
Dr. Hynek's ". . . Good Book About UFOs"[2]

The UFO Experience: A Scientific Inquiry [13] is a serious book and probably the most influential book written on this subject to date. It's hard to imagine how there could be a more qualified author. Dr. Hynek was an astronomer with impeccable credentials and consultant to the U.S. Air Force on the matter of UFO's for more than twenty years (Project Blue Book.) There is no doubt that Dr. Hynek thoroughly understood the scientific method and did his best to apply it to this subject. Unfortunately, all of the raw data that he had to work with were anecdotal in nature, and getting the Air Force to behave scientifically proved to be an uphill battle. Nevertheless, he did a commendable job of winnowing the wheat from the chaff and looking for the credible tidbits that could be used to establish prototype cases.

Dr. Hynek established six categories of sightings into which all of the Project Blue Book data are assigned. These are arranged in what was called "an increasing order of strangeness." **Table I** lists these categories and a brief definition of each.

---

[2] [13] p. vii

Table I
UFO Sighting Categories

1. NOCTURNAL LIGHTS (NL) — Lights in the night sky, the least strange and most frequently reported event.

2. DAYLIGHT DISCS (DD) — Observations made during daylight hours of oval or disc shaped objects.

3. RADAR-VISUAL (RV) — Instances involving both radar and visual mutual confirmation.

4. CLOSE ENCOUNTERS OF THE FIRST KIND (CE-I) — Objects or very brilliant lights close to the observer, usually less than 500 feet.

5. CLOSE ENCOUNTERS OF THE SECOND KIND (CE-II) — The reported UFO leaves a visible record of its presence.

6. CLOSE ENCOUNTERS OF THE THIRD KIND (CE-III) — UFO reports in which animated creatures are present.

Some 12,000-plus reports are in the Blue Book files, of which over 700 are listed as unknown. Dr. Hynek sifted through these to select the most credible as prototypes. He tried to choose those that had received a thorough investigation, and included many in which he had been the principal investigator. He favored those in which there had been more than one witness, and those where the witnesses held responsible, professional positions or had a substantial educational background. Finally, he narrowed the data down to about a dozen or so prototype cases for each of the established categories, a total of 80 all together. These prototypes are summarized in Appendix I of his book. In addition, the text of the book includes more complete descriptions of some of the most compelling reports. It is difficult to read these selected reports without becoming persuaded that some real physical phenomena are being observed. Here is how Dr. Hynek expresses *his* conviction on this point:

> "There exists a phenomenon, described by the contents of UFO reports (as defined here), that is worthy of systematic, rigorous study. The extent of such a study must be determined by the degree to which the phenomenon is deemed to be a challenge to the human mind and to which it can be considered potentially productive in contributing to the enlightenment and progress of mankind."[3]

Much has happened since Dr. Hynek's book was copyrighted in 1972, but the basic question remains unanswered and much of the present book is dedicated to speculation on this

---

[3] [13] p. 242

question. In these speculations I have attempted to maintain the same high standard established by Dr. Hynek, staying as close as possible to science as we currently understand it. Using Dr. Hynek's prototypes, I have created a list of the most commonly mentioned characteristics. These are listed in **Table II.** It is worth taking the time to become familiar with this table because it is one of the key features of this book and will be referred to repeatedly.

Table II
Commonly Mentioned Characteristics

1. The subject was often said to be oval, oblong, cigar shaped, a disc, or like a saucer, but sometimes circular.

2. Bulges were noted which were sometimes described as a dome on top of the saucer.

3. The subject was sometimes described as rotating, and frequently said to have a wobbling or rocking motion when moving slowly or hovering.

4. The subject was often said to be multicolored, with red, orange and green being mentioned the most, and as changing from one color to another or pulsing.

5. The subject was often said to be luminous or to have a fluorescent quality sometimes described as shimmering, shiny, or having metallic iridescence.

6. The subject was frequently described as fuzzy at the edges or having an indistinct outline, but not always.

7. The subject was most often described as completely silent.

8. The subject was described as capable of incredible accelerations and very high velocities.

9. The subject was described as capable of sharp changes of direction at high speeds and of attaining very steep ascent or descent, often in a stepped manner.

10. The subject was often said to appear in multiples sometimes described as the mother ship plus progeny or as splitting apart and rejoining or simply disappearing.

11. The subject was often (but not always) sighted over or near bodies of water.

12. In close encounters, the subject was frequently said to have killed the engine of motor vehicles, or to have left burned or scorched marks on the ground.

Chapter 2
A Close Encounter of the First Kind

I have known Gordon Chism for ten years and I know him to be a perceptive, intelligent and honest person. Gordon is a firm believer that at least some UFOs are the result of an alien presence. If seeing is believing then Gordon has good reason to believe as he does. Many years ago he had an experience, which I have persuaded him to relate publicly for the first time. In addition, he has made a study of the literature on this subject, and he is the one responsible for giving me Dr. Hynek's book to read. Here then, is Gordon Chism in his own words as recorded on tape after a recent dinner party:

"It was the summer of 1958; we were about 70 miles east of Reno near Fallon, Nevada. Just outside of Fallon is Stillwater, Nevada which is the Carson sink. We were teenagers, eighteen years old and seniors in high school. With me were my best friends from high school, Bill Rose and Ken Taber. My brother Will was also along on this expedition so there were four of us. We stayed at Ken Taber's father's duck hunting cabin which was Spartan but livable. We would go out there on weekends and carp hunt with bows and arrows. We had this 1936 Plymouth that we had taken the body off of and got it down to the barest of essentials with absolutely nothing but the engine, the cowl, the steering wheel and the front seat. In the back we bolted a sheet of plywood so we could carry stuff on it.

Anyway, we were out carp hunting and it got to be dusk so we decided we'd better get going back to the cabin because we had no headlights. We were all worn out anyway. So we got up on the Division Road — a road that goes to the center of the Carson Sink and is about six or seven feet above the rest of the swamp. On one side of this road was the National Wildlife Conservancy and on the other side were the duck hunting cabins and club. We only had about three miles to go back to the cabin on Division Road. It was dark. The light was fading. Bill Rose, who was sitting on the back, said "You had better speed up we have a car on the road behind us." We all looked back and sure enough on the road behind us was this bright light about 100 to 150 yards behind us. So we started going a little bit faster when Bill Rose said "Don't bother the car just went up in the air." We stopped and looked back and sure enough this thing was rising. It went up about 100 feet or 150 feet and then it just stood there. There was about a ten mile an hour breeze blowing but it was just nailed solid against the background. It would glow very brightly and then get dim. It was sort of oval in shape and about 20 feet in diameter. None of us could say that it was a solid metallic object with rivets, doors and hatches or hard edges. It was just this glowing sort of plasma thing. Sometimes it appeared to have a rotary motion going on around its perimeter. It would get dim and then bright. It would stay bright for about 10 seconds and then take about 15 seconds to get dim again. It kept repeating this pattern. Our jaws dropped. We were transfixed by this thing. We were all into physics, science, cars, and airplanes. This wasn't an airplane! It was just there! We didn't know what it was! We watched it for about ten minutes. Then it started to go to the south of us about 200 yards away. It was getting dark so we took our swamp buggy back to the cabin, got into Bill's station wagon and went back out to watch this thing again. We drove back out to where we had been, but it wasn't there! It was gone and we started to make jokes about UFOs etc. Since it was gone we decided to go back and make dinner. We started back when my brother Will said, "Nope there it is!"

There was this line of poplar trees and this thing had landed behind them. Then it rose up about 20 feet above the top of the trees and did its rhythmic glowing thing. Then from the ground up comes this bright red sphere about two or three feet in diameter. It came off the ground and went into this thing! Then it stared moving again and it started going toward the cabin. So we followed it back to the cabin and parked the car. We all got up on the roof to watch. So there's the four of us on the roof watching this thing move along, and it got to be about a half mile away from us. It was moving mechanically, just slowly moving. It would glow brightly and then dim. There was a haze cloud cover. It would light the haze above it and the ground below it. Repeatedly it would get dim, then glow brightly and light up the haze above.

We're sitting on the roof of the cabin watching. Nobody said a word. We were mesmerized. We watched for, oh, a good half hour maybe as much as 45 minutes. While we watched it moved south of us then turned right moving along parallel to us. Then it stopped about a half mile away. It was stationary, absolutely stationary, against the background. All of us were transfixed watching it. It was there *and then it was right on top of us!* Instantaneously! It didn't seem to accelerate. It was there and then it was here! We all leaped off the roof, ending up on the ground. When we got up it was gone! This thing had set us up and knocked us off. It was not only intelligently controlled; it had a frat-house sense of humor!

We spent our whole night peering out windows, going outside and looking around then coming back inside. It never came back. It was a very restless night. When we got up in the morning we went to a gas station and asked if they had seen a strange light last night. The gas station man gave us the "Oh yeah, what are you guys drinking." That was it. We didn't see anything more. We did call the Fallon Naval Air Station and asked if there was anything in their area that caught their attention last night. They said, "Nope the scopes were clear."

Several years later *Look Magazine* — I think it was *Look*, but it might have been *Colliers* or *Life* — had an article about the same kind of thing we had seen! They had pictures of it. It was somewhere in Michigan. Over a hundred people had watched this thing for approximately an hour. The description was very much like what we had seen. I got three or four copies of the magazines and gave them to my friends with "Hey we're not crazy after all."

But the curious thing is that my friends — even my brother — said, "Oh well, we saw a light in the sky." Recently, (over forty years later) I told this story again. My brother was present and he said "Yeah you're telling the story right, but it was just a light, just a light in the sky." It seems none of them were really intrigued about it except me. It was as if they didn't want it on their record or something. But I was fascinated and it shook me up because I was very much a skeptic about that sort of thing. I thought I knew everything and obviously I didn't."

I thought it would be interesting to see if I could locate a back issue of the magazine that Gordan mentions above and a little searching turned up an issue of Life Magazine *that describes the events in Michigan as well as others. The article includes the following quote:*

"Some are greenish and iridescent, like the mystery thing that swooped over Perth, Australia several weeks ago. . . Others are football-shaped and glow with pulsating lights. Last week the manifestations seemed almost to have reached the proportions of an invasion. Near Ann Arbor, Mich. 52 witnesses, including a dozen policemen, saw five strange objects hovering over a swamp. . . The next day a glowing thing floated over a small college in Hillsdale, Mich. and was sighted by 87 students, an assistant dean and the local civil defense director. Whatever the explanation of the peculiar phenomena — seen and described similarly by so many — *something* surely was in the air."[4]

---

[4] O'Neil, Paul "A Well-Witnessed 'Invasion' by Something" *Life*, April 1, 1966 p. 25 Also see [14] p. 200

etching   Sun Strokes   R. L. Mason

Chapter 3
Natural Speculations

It seems possible to me that what is being observed is the fortuitous concourse of a number of rare, but natural, events.

Swamp Gas

One candidate for such a rare but natural phenomenon is swamp gas. Swamp gas consists of a mix of natural gases generated by the decomposition of vegetation and other organic materials and is usually found in marshy conditions. Occasionally a concentration of such gas, which is predominately methane, appears to spontaneously fluoresce. This is most often witnessed when ambient lighting is low.

Dr. Hynek was dispatched by the Air Force to investigate the events in Michigan mentioned in the Life Magazine article. He suggested swamp gas as a possible explanation for what was observed there. He was roundly and soundly condemned for this by witnesses and the press. I discussed with Gordon Chism the likelihood of swamp gas being responsible for what he saw, but he wouldn't buy it. As he points out in his narrative, there was a breeze blowing at the time yet the object often appeared completely stationary, and it turns out the sightings in Michigan were also on a windy evening.

Thought Experiment #1

Here is a little thought experiment I devised for myself to try and explain some of the items in **Table II**, the list of

Commonly Mentioned Characteristics. Imagine that you are sitting in the corner of a small, completely dark room. The ceiling of this room has a large circular hole in it. You cannot see this hole. The hole connects the room with a similar room just above the one you are in. In the room above, and just over your head, is a laser level of the kind used in construction for casting a level line. This device is very close to the floor of the room above and casts a line across the hole that shows up on the opposite wall, but you can't see this because you are below.

Now, imagine that someone in the room above pours a thin stream of a transparent viscous liquid, like motor oil, down through the hole. What you can see from your position below is just a red dot where the liquid passes through the plane of the laser light line. Furthermore, because you view this dot at an angle, the dot appears to be an ellipse or disc shaped. Also, because there will be some scattering and diffraction as the laser light strikes the stream of liquid, there may appear to be bulges at the center of the dot.

Now, the person above moves the stream rapidly in a random fashion. The dot will appear to move, making sharp changes of direction and exhibiting extreme accelerations and high velocities. Sometimes the line of the stream will approach the laser plane at an angle. This will increase the apparent speed of the dot and cause it to elongate even further. Now the person above rotates the device from which he pours the stream, imparting a twist and causing the dot to appear to rotate or even break into several dots as the stream does likewise. Finally, imagine that the laser line is raised or lowered causing the dot to ascend and descend, sometimes in a stepped fashion.

It seems to me that this experiment may be analogous to something that actually happens in nature and may explain, if only through the various geometric relationships involved, some of the items in **Table II,** particularly items 1, 2, 3, 8, 9 and 10. The following three sections explore this possibility.

Particle Accelerators

Over the years I have acquired a superficial knowledge of the workings of particle accelerators. It comes partly from a paper I once wrote for a high school physics class, and partly from association with friends and relatives who worked at the Stanford Linear Accelerator (SLAC), but mostly from two years of employment as an engineer with the Radiation Division of Varian Associates in Palo Alto where accelerators were manufactured. Particle accelerators come in two basic configurations: circular (cyclotrons and synchrotrons) and straight (linear). The accelerators built by Varian were linear and were used for research at various laboratories and educational institutions, or to produce x-rays for medical purposes such as zapping cancer tumors.

Briefly, a linear accelerator is a segmented tube of individual cavities with a gun at one end and a target at the other. The gun is heated up and boils off a stream of charged particles such as electrons, which are aimed down the tube by a parabolic reflector and a collimator. The collimator is simply a disc or series of discs with a small, centered hole. It sizes and lines up the particle stream with the tube. Once in the tube the particles "surf" on a traveling radio frequency wave whose phase velocity is increased as it travels down the tube, thus accelerating the particles. The longer the tube, the higher the velocities and energies achieved. At two miles, SLAC is the longest linear accelerator in the world. The accelerators built by Varian were much smaller versions.

The tube is usually surrounded by doughnut shaped electric magnets that squeeze and concentrate the particles. When the particles impact the target, x-rays are produced which can be used for medical or other purposes. When energies are high enough, the particles blast apart the atoms of the target and particles of all sorts fly in different directions. The paths of these particles are recorded on film through the use of a device called a cloud or bubble chamber. By analyzing the paths of the particles, a researcher can tell much about their nature.

How About the Sun?

All of this usually takes place in a laboratory of some kind, but could it happen in nature? It is my thought that something similar could happen on the Sun. We know the Sun is a fusion reactor. We can produce fusion reactions here on earth. They are called hydrogen bombs. The Sun is one big continuous hydrogen bomb.

I read with interest an article entitled "Here Comes the Sun" by Dana Mackensie[5]. Solar flares that originate from Sunspots can spew-forth huge amounts of hot plasma, or streams of high-energy charged particles. There are eruptions on the Sun every day of a common garden variety, and they send forth ionized particles and electrons, some of which are aimed at Earth. These are deflected by Earth's protective magnetic field. They are routed along magnetic lines of flux to the pole areas where they may be observed as auroras. There

---

[5] *Discover*, May, 2004 p. 63

are, however, magnetic storms that take place on the Sun and these can be seen as Sunspots which travel across the face of the Sun, taking about ten to twelve days from side to side. The spots are the source of solar flares which send out plasmas and high-energy particles which are much more powerful than normal Sun activity. These solar storms are periodic with occasions of high activity and periods of relative calm. The time between periods of high frequency is on average about 11 years. The latest high activity period, at this writing, started in 1998, peaked about 2001, and ended in 2006.

Here is a quote from the same source that caught my eye:

"Nearly all the activity scientists see on the Sun's surface — Sunspots, flares, coronal mass ejections — is governed by mysterious twists and turns in the [magnetic] field."

The *Discover* article concludes with the following statement:

"Still, the list of what scientists don't understand about the Sun is daunting. For example, what drives the 11 year sunspot cycle, what makes the corona so hot, and how big do solar flares get?"

It is my conjecture that there may be some mechanism or combination of circumstances on the Sun that produces a particle stream that is focused and cohesive in the same way as one produced in a lab here on Earth, but immensely more powerful. Such a stream could possibly be associated with Sunspots and could, under the right conditions, send a stream directly at Earth. However, even though this stream may be tightly focused and cohesive, it would probably not be a perfectly straight line. It would have squiggles and twists and maybe even rotate as the conditions that produced it varied on the Sun's surface. Such a stream, with sufficient energy, would be able to penetrate the Earth's magnetic field without being deflected to the polar areas as normal charged particle activity is. In fact, it would produce a "storm" in the Earth's magnetic field that would play havoc with electrical equipment. It might even induce a current in automotive electric circuitry that would be of opposite polarity to the system's normal orientation and, thus, cause the engine to stop.

If, on the way to the Earth's surface, such a charged particle stream happened to pass through a layer of gas that was trapped or concentrated in some manner, perhaps by a temperature inversion layer, it could possibly light up the gas and appear as a fairly discrete object. If the apparent spontaneous fluorescence of swamp gas mentioned earlier in this chapter were actually triggered by a charged particle stream then it could appear stationary even in the presence of breezy conditions. Such an "object" could easily satisfy all the characteristics listed in **Table II**, and would naturally reproduce the various geometric relationships of **Thought Experiment # 1**.

The operating principle involved would be the same as that which enables fluorescent or neon lighting. An electrical charge passing through an appropriate gas produces light. An apparent object that emits a fluorescent, multicolored, pulsing light, and is slightly fuzzy or indistinct, and moves erratically is explained most easily. A daylight disc or radar contact is not as obvious but I believe, still possible.

Testing for a Correlation

The best thing about this hypothesis is that it is testable. If true, then there ought to be a correlation between the frequency of UFO sightings and the Sunspot cycle. Hynek does a considerable amount of grousing about the very poor data collection of the Air Force, and how they could have done so much better. There was one piece of data, however, that was always collected and is probably very reliable, namely the date of the sighting.

I obtained a plot of solar Sunspot activity from the National Geophysical Data Center web site[6] for the period covered by Hynek's book, and plotted the dates of the 80 prototype cases against it. There was no obvious correlation, but with only 80 data points it was probably too much to expect. Back on the Internet I found a summary of Project Blue Book cases on the Freedom of Information web site[7], and lo and behold, this included a full list of sightings for 1947-1969. I plotted this list against the Sunspot cycle, but again, no obvious correlation. Next, I found the National UFO Reporting Center (NUFORC)[8] with data right up to the present day, but again, nothing definitive. Finally, I decided why not just go to GOOGLE and enter "UFO Sunspot correlation?" I did this, and up popped a fellow named Larry Hatch[9] who had "been there, done that" and, alas, found no correlation. I still think the charged particle beam concept has merit, but I am forced to one of four possible conclusions:

---

[6] http://www.ngdc.noaa.gov/stp/CDROM/solar_variability.html

[7] http://www.reflectionsinthenight.com/blue_book_1.htm

[8] http://www.nuforc.org/webreports/ndxevent.html

[9] http://www.rense.com/general3/sunufo.htm

1. The lists of UFO sightings include so much irrelevant data that any possible correlation is masked; or

2. Beams of charged particles come from the Sun, but have no connection to Sunspots; or

3. Charged Particles arriving from the Sun are too defuse to generate an apparent object; or

4. A charged particle stream or beam comes from *some other source.*

Chapter 4
Domestic Speculations

Robert T. Jones, Aerodynamicist

When I was a youth, in the years immediately proceeding and following adolescence, one of my best buddies was Ed Jones. Ed and his two sisters lived three doors from me in Palo Alto and I spent a lot of time at their house. Their father was Robert T. Jones, the famous aerodynamicist. I didn't realize how famous he was at the time. To me he was just their father and he worked for NASA, although the kids were fond of pointing out that *their* father had invented the swept back wing. Hanging out at the Jones' house was a real education. Ed was a few years older than I and he had an old Model T. Ford truck that we were continually tearing apart and putting back together. In addition, we built countless model airplanes. In this latter endeavor we received the most expert instruction you could possibly imagine. There were also lessons in optics and electronics. Most of my instruction came to me second hand from Ed, but occasionally I got it straight from "R.T." himself.

R.T. had a knack for making the strangest looking contraptions fly, and fly well. One in particular sticks in my memory. It was a flying wing, but more like *half* a flying wing. It was almost all wing and it was straight, but it flew on a bias. In flight, one end would lead by quite a bit and the other end would trail. Looking at it made you scratch your head and think, "How could such a configuration possibly fly?" but it did. R.T. called it an "oblique flying wing." Many years later, after he retired from NASA and was associated with Stanford University, he spent time developing this concept further. What eventually emerged was a fairly complete design for a Mach 1.6 aircraft that could seat 440 passengers inside a wing with a span of 400 feet. Of course, you don't see anything like that flying today. It was just too strange for the public to accept. As R.T. himself once wrote "Artifacts created by humans show a nearly irresistible tendency for bilateral symmetry."[10] If one of these wings were to appear in the sky without warning you would immediately be struck by the almost "unearthly" quality of its appearance.

There probably are aircraft of various types and configurations under development at secret facilities, and no doubt they contributed from time to time to the total inventory of UFO sightings. The recent unveiling of the stealth bomber and fighter is an example of this kind of secret development.

Area 51

Among UFO enthusiasts are those that frequent the small town of Rachel, Nevada, located near Area 51 and the "secret" base at Groom Lake. There you can sign up for a guided evening trek up into the Nellis Range over looking the base. I have never done this, but I understand that on Wednesday nights at a certain location, it can be a very interesting experience, or at least it used to be. The most common description of what was seen involves a ball of brilliant light that changed altitude in a stepped fashion, exhibited terrific accelerations,

---

[10] Robert Thomas Jones, May 28, 1910 — August 11, 1999 by Walter G. Vincenti / Biographical Memoirs, National Academy of Sciences

and made sharp turns at very high velocities. This has all the ear marks of a beam of some kind directed from either above or below — probably below. But how could such a beam be made to exhibit a bright ball of light without hitting a target or passing through a layer of gas and thus causing fluorescence? I don't know, but I am addicted to speculating about this kind of thing.

If there was a device similar to the linear accelerators described in the previous chapter that produced a particle beam of some kind, maybe there is a way to run something up and down that beam. Suppose this device produced a beam of protons or positive ions and accelerated them to high velocity. And suppose there was a way of superimposing a wave function on this beam. I envision this as looking like stop-and-go traffic on a freeway when viewed from above. And further, suppose there is a way of controlling the phase velocity of this wave. This gets me back to R.T. Jones again.

In the early days of television, when I was somewhere in the range of ten to twelve years old, I happened to walk into the Jones's living room one day and noticed what I thought was a TV. Not everybody had one in those days, and I commented on its presence. R.T. was nearby and he informed me that what I was looking at was actually an oscilloscope.

"What's an oscilloscope?" I asked, stumbling over the pronunciation.

"Do you know what oscillation is?"

"Uh huh", I replied with a nod, not wishing to appear ignorant.

He looked at me for just a moment and an almost imperceptible smile crept over his face.

"Well oscillation is when something moves back and forth between two limits like the pendulum on a clock."

I remember being embarrassed that he had seen through me so easily. He then turned on the scope and gave me a lesson in its use. The image that sticks in my head to this day was the standing sign wave that he produced on the screen. He made it proceed to the right slowly at first then gradually faster until the wave was just a blur. He then slowed it down again until it was motionless again, all the while explaining about frequency and phase velocity. Anybody who has worked with an oscilloscope has seen what I have just described. Much later, when I worked for the Radiation Division of Varian Associates, I was impressed with the fact that the phase velocity of a radio frequency signal can be used to move particles down the length of an accelerator's tube, and accelerate them in the process.

So, what if we pass a positive beam through a small synchrotron-like device that has electrons whirling around in it, and we pass the beam through the synchrotron along its axis of rotation. Would there be a way of hanging a ring of electrons on the positively charged beam? And could this ring be moved up and down the beam or be held steady at some location by adjusting the phase velocity? And since the ring of electrons would give off what is called synchrotron radiation from being constantly held to a circular path, would this electromagnetic radiation be in the visible portion of the spectrum? There are a lot of questions here and not many answers.

After thinking about this a little more, I realized the synchrotron is not necessary. All you need is a small linear accelerator arranged perpendicular to the larger positive beam. This linac would put out a stream of electrons at just the right velocity and distance from the center of the positive beam so that they go into a stable orbit around the stream of positive particles. The phase velocity of the positive beam could be adjusted so the positive peak of a standing wave (zero phase velocity) was at the intersection of the two beams while the orbit was being established. Once established, the phase velocity of the positive or carrier beam could be increased, sending the ring on its way.

After writing the above I decided to see if anybody else had come up with the idea of linking particle beams and UFOs and this led me to the web site authored by Tom Mahood entitled Bluefire.[11] From his site, I gathered that Tom was at one time an Area 51 groupie, but has since reformed and is now a physicist. This combination of attributes provides him with a unique perspective on the question of UFO's sighted over Area 51. Tom speculates that a particle beam aimed up into the atmosphere would, depending upon the initial energy and velocity, produce a ball of plasma at some altitude, and he provides some impressive mathematics to back this up. Coming out of the accelerator at high velocity the beam would initially shoulder aside molecules of air, but would gradually attenuate to the point where it would eventually collide with these molecules and dump its remaining energy creating a ball of

---

[11] http://www.serve.com/mahood/bluefire.htm

plasma. A good deal of the resulting electromagnetic radiation would be in the visual portion of the spectrum but it would also show up as a false bogie on radar screens, and he suggests that is where the military interest lies. Mahood notes a paragraph from David Darlington's book, *Area 51 — The Dreamland Chronicles*[06]. Darlington, quotes Mark Farmer and because it is such a descriptive passage, I will quote it also:

> "I've seen two of them out here," Farmer divulged. "One was a light that kept bouncing around and then just went away. The other was colored, floating, glowing orb that popped up behind the jumbled mountains south of Groom Lake. It went straight up, then started jerking around and wobbling up and down — at times making right-angled, or greater than right-angled, turns then sitting still in a rock-hard hover. It became distorted when it moved part of it lagged behind the main object, then the trailing edge would catch up. I had a Celestron twelve-hundred-millimeter telescope, and I watched it for an hour and forty-five minutes. It wasn't quite round; it was sort of squashed, and shimmering the whole time as if it were surrounded by some kind of field. It was crimson on top, blue-green on the bottom, and gold in the middle. I have no idea what it was."[12]

After reading his on-line essay "Particle Beams and Saucer Dreams,"[13] I had the feeling that Mahood is probably closer to established science than I am, but who knows what is yet to be

---

[12] [06] p. 237

[13] http://www.serve.com/mahood/probeams.htm

established? I'll touch on this kind of thing again in the next chapter, but from a different point of view.

Historical Coincidence?

The first particle accelerator, a cyclotron, was built in the early 1930s, but they really didn't proliferate until the invention of the synchrotron and the linear accelerator in the mid 1940s. I think it is interesting to note that the modern era of UFOs is generally considered to begin in 1947. As Tom Mahood speculates, the military interest in accelerators probably has to do with their ability to create a false radar target. Just how much development and distribution the resulting device has received is unknown, but I get suspicious whenever I read about UFOs being sighted in conjunction with large military maneuvers or naval exercises.

Chapter 5
Alien Speculations

There is an interesting "if-then" logic to the alien explanation for UFOs. *If*, in fact, they are here, *then* they must be more advanced technologically than we are. This assumes visitors from a different star system, a trip we are currently incapable of making. How advanced would they be? Unknowable, but the possibility of an advanced technological civilization certainly exists. The age of our solar system is currently estimated to be approximately 4.6 billion years and the Universe, 13.7 billion years. So there has been plenty of time for such a civilization to develop and precede us. Here is how Dr. Hynek summarizes the concept in his book:

> "We work in the brilliant spotlight of the present, only dimly conscious of the penumbra of the past and quite unable to illuminate the darkness of the future. Let us imagine for a moment a covered wagon train of not much more than a century ago, winding its long journey to the west. It is encamped for the night, its wagons in a circle, sentries posted, and the travelers gathered about a campfire for warmth and cheer. Someone speaks of the future, but he speaks, as he must, with the words and concepts of his day. But even were he inspired by some kindly muse of the future to speak of making their entire journey in a matter of hours, flying through the air, and of watching scenes by television and hearing voices speaking on

another continent, this gifted one could not have put into words a glimmer of how these wondrous things might be accomplished. The vocabulary for such descriptions electrons, transistors, integrated circuits, jet engines the jargon vehicle of technical communications would not exist for yet a century. He would be helplessly incoherent for want of words as vehicles for his thoughts. . .”

I would interject at this point that Dr. Hynek certainly has words as vehicles for *his* thoughts. He continues. . .

“Even if we limit our thinking to the billions of stars in our galaxy alone, we know that our galaxy was in existence for billions of years before our sun appeared. Thus the stage was set long ago for the possibility of civilizations as greatly advanced beyond us as we are beyond mice.”[14]

I don't see how one can argue with that. The logic is inescapable.

The Search for Extraterrestrial Intelligence (SETI)

The SETI era, for the most part, began with Frank Drake, although at least one paper was published expressing the concept slightly before Drake was able to get into action.[15] In 1960 Drake aimed the 85-foot radio telescope in Green Bank, Virginia at the Sun-like stars Tau Ceti and Epsilon Eridani. Called Project Ozma, the effort continued for sixty

---

[14] [13] pp. 262 & 263

[15] G. Coccni & P. Morrison “Searching for Interstellar Communications” *Nature*, 1959

days without producing any definitive results. It did, however, generate a lot of interest both inside and outside the scientific community. Since that time, a number of different efforts have been made, each increasing in scope and sophistication. The search continues, primarily through the use of radio telescopes, although other techniques are being investigated.

At this writing, the latest to come on line is the Allen Telescope Array in the Cascade Mountains of Northern California. The Array consists of 42 dishes, each 20 feet in diameter, and is sponsored by Microsoft co-founder, Paul Allen. It is probably the most technically advanced facility to date.

The Drake Equation

It has become the second most famous equation in the world after Einstein's $E = mc^2$ and is generally known as "Drake's Equation" or "The Drake Equation." It was first penned by Frank Drake in preparation for what has become known as the Green Bank Conference of 1961. It took a while for it to gain the prominence that it now enjoys and, in fact, it is not mentioned in Dr. Hynek's book which was copyrighted approximately ten years later. However, it seems prudent that any update of our knowledge on the subject at hand should include the most current thinking in this regard.

The equation reads as follows:

$$N = R^* f_p n_e f_l f_i f_c L$$

where:

N = the number of civilizations capable of radio communication

R* = the number of life-friendly stars born each year in our galaxy

f_p = **the fraction of suitable stars that have planets**

n_e = **the number of habitable planets around each star**

f_l = **the fraction of habitable planets that have life**

f_i = **the fraction of habitable planets that develop intelligent life**

f_c = **the fraction of intelligent species that develop radio technology**

L = the average time span that civilizations transmit detectable signals

The only term we have a handle on in this equation is **R*** which was originally set at ten, and some progress has been made on the second term since 1961; but all the rest are largely conjecture based on only one data point, namely our existence here on Earth. The original values assigned to the variables in 1961 produced an answer in the thousands, and Seth Shostak, spokesperson of the SETI Institute, feels that even though some opinions have changed in light of new information, "thousands" is still a reasonable order of magnitude.[16]

There exists a slightly different version of the equation which reads as follows:

$$N = N_* f_p n_e f_l f_i f_c f_L$$

---

[16] Seth Shostak "Drakes Brave Guess" *Discover* May, 2006

where:

N_* = the number of stars in the Milky Way Galaxy

and

f_L = the fraction of a planetary lifetime for which a technical civilization exists

All the other variables are the same and the results are comparable, but perhaps it is a little easier to follow. Of course, if any term in the equation is zero then **N** must be zero, but we know that **N** is at least one. So what we have here is a situation where the compounded product of all the various probabilities is probably quite small, but fortunately some of the multipliers are very large.

Perhaps the real value of the Drake Equation is the way it organizes our thinking on the subject. One version or the other has been printed repeatedly in books, magazines and professional journals, as well as on T-shirts and coffee mugs. It has been featured on TV, including Carl Sagan's *Cosmos* series. There are even web sites where you can go and plug in your own best guesses for the variables and **N** will be calculated for you.[17] I did this and got a value of 1500. If the value of **N** *is* in the thousands then there should be at least one such civilization within approximately 1000 light-years of Earth. That gives you a rough idea of how large the Milky Way Galaxy is. 1000 light-years may seem like an impossibly large distance, but consider that the Galaxy is estimated to be 100,000 light-years across.

---

[17] http://www.activemind.com/Mysterious/Topics/SETI/drake_equation.html

The Implications of Quantum Physics

Dr. Hynek, in the concluding sentence of his book, writes:

> "When the long awaited solution to the UFO problem comes, I believe that it will prove to be not merely the next small step in the march of science but a mighty and totally unexpected quantum jump."[18]

His use of the word "quantum" may turn out to be fortuitous *or did he do it on purpose?* Since the publication of Hynek's book in the 1970's there have been a number of developments in the field of quantum physics which may have some bearing on the UFO question. The phenomenon known as *quantum entanglement* holds that when two or more particles are put into the same state they become "entangled" and forever afterward mirror each other no matter how far apart they are separated. Moreover, this communication is instantaneous. The mathematics governing such action was first pointed out by Einstein *et al.* in 1935[19], but he refused to believe it was possible, calling it "spooky action at a distance." He proposed that the theory must be incomplete.

However, subsequent developments have supported the validity of entanglement, and it has recently been verified experimentally on numerous occasions. This would seem to open up the possibility of a number of practical applications. Those mentioned so far include unbreakable encryption, parallel processing by quantum computers, and teleportation.

---

[18] [13] p. 264

[19] Einstein, Podolsky & Rosen "Can Quantum-Mechanical Description of Physical Reality Be Considered Complete? " *Physical Review* 41,777, May 15, 1935

The last of these would seem to border on science fiction but, in fact, teleportation has already been achieved on a very small scale (subatomic particles) over short distances.

Even though I haven't seen it mentioned, *remote sensing* would also seem to be a possibility. How would that work? It is next to impossible to know or even guess how a technology that may be centuries, millennia, or even much larger time periods, in advance of our own, would work. As Arthur C. Clark states, "Any sufficiently advanced technology will be indistinguishable from magic." Nevertheless, I believe we must, at least, try. Extrapolating on current human technology, I imagine a device like a cyclotron or synchrotron accelerator in which entangled particles such as electrons or protons are inserted and accelerated to velocities approaching the speed of light. Then some portion of them is spun off on a tangent pointed at a distant target. This would constitute a beam of charged particles. The remaining particles are maintained at velocity in a storage ring something like those in use today at various high-energy labs. When the targeted particles reach their destination they interact with whatever is there and that interaction is mirrored by their entangled "brethren" in the storage ring at home.

Maybe it would work something like the speculation in the previous chapter. The entangled particles are put through a linear accelerator, then a beam splitter of some kind divides them sending half to a storage ring and putting the other half in orbit around a carrier beam of opposite charge that has wave characteristics. Care would be taken to achieve an exact match in the conditions the two halves experience, then the traveling half is sent on its way in an annular configuration, somewhat

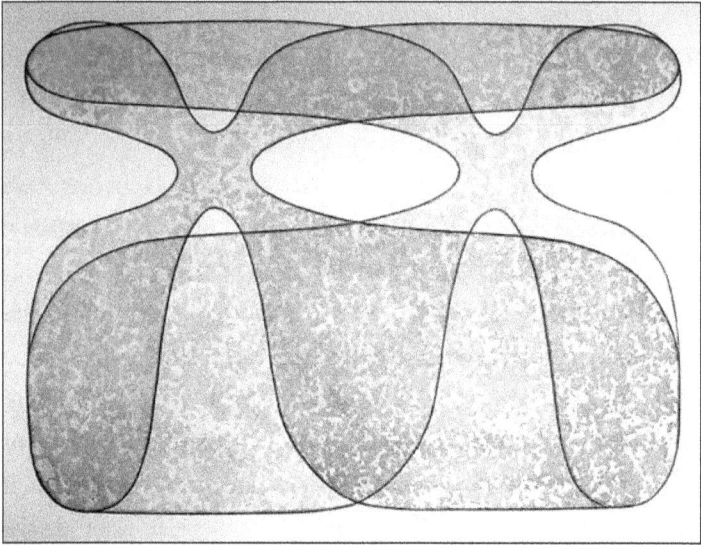

etching   Entanglement   R. L. Mason

like blowing a smoke ring. One of the disadvantages of using a charged particle beam as a sensing device is it is easily deflected by magnetic fields, but a beam organized, as described above, could present itself as largely neutral and would be expected to travel in a straight line. Since a particle *must* have a charge in order to be accelerated, the neutralization will have to be accomplished *after* the acceleration is complete.

A sufficiently advanced technology would have some way of monitoring what happens to the particles in the storage ring and thus could sense the nature of the distant target. Such a monitor might make use of synchrotron emission (electromagnetic energy that radiates from such a ring as a result of the particle's angular acceleration) for this purpose. This electromagnetic radiation might also be used at the target end for sensing the environment there.

This is all wildly speculative, but maybe an advanced alien culture in some distant star system many light-years from Earth could remain at home while monitoring developments in our world. The energy requirements to operate such a system would be impressive, but still be considerably less than sending an actual physical space craft, not to mention the problems involved with sustaining and protecting alien beings during such a voyage. A charged particle beam could be accelerated close to the speed of light fairly easily and sent on its way while the aliens stay in the comfort of their own world. They might still have to wait considerable periods of time for the beam to reach it's intended destination, because even at velocities close to the speed of light the trip could take many years, but once the destination was reached they would get instant results back home.

Active and Passive Sensing Systems

The human eye is an example of a passive sensing system. It gathers ambient light and transmits it as data to the brain. Hearing is also a passive sensing system. Radar and sonar are examples of active sensing systems. A signal is sent out and that portion that bounces back to the receiver is sensed. If a person picks up a flashlight in the dark and shines it in order to see, that person becomes an active sensing system. Electromagnetic radiation in the form of visible light leaves the flashlight hits the target and bounces back to be collected by the eye. Almost all astronomy conducted by humans has been passive, although radar signals have been bounced off the Moon and have also been used to map the surface of planets from spaceships in orbit. A telescope is a passive sensing system, even though it greatly enhances the ability of the human eye to collect electromagnetic radiation. What I am suggesting is that alien beings on distant worlds, with sufficiently advanced technology, might use an active sensing system to do astronomy. We know how to do this on a small scale. An electron microscope is an example of an active sensing system using a beam of electrons to obtain resolutions much finer than can be achieved with electromagnetic radiation (light). The disadvantage of an active sensing system for great distances, as in astronomy, is the signal must make a round trip and the astronomer must wait twice as long for results. However, by utilizing the instantaneous communication of entangled particles the aliens could get around this problem.

Would this alien monitoring process be apparent to us here on Earth? Under most conditions a cohesive, high-energy beam striking Earth might not be visible to the naked eye. If, however, it were to pass through a gas such as methane upon its arrival it could cause the gas to fluoresce, although it would have to carry a charge for that to happen. Such fluorescence might act as a natural monitor revealing the organized nature of the beam. Maybe, instead of fluorescence we would see a plasma display as described by Tom Mahood (page 29). This would be especially true in low ambient light conditions. Under these conditions one might see a kind of hologram of the storage ring back on the alien world. Thus, the storage ring on the alien planet becomes a virtual spaceship. It may even be possible to "steer" or control the traveling half of the entangled particles by manipulating those in the storage ring on the home planet. In other words, they could "fly" their virtual spaceship to various locations of interest and all the while staying physically at home. This would be roughly analogous to the pilotless drone aircraft now in use by the U.S. military. The operator remains safe and secure back at the base and flies the drone out over hostile territory. On those rare occasions when conditions are just right to reveal their virtual spaceship to the human eye, it seems quite likely that most of the commonly mentioned characteristics of UFO sightings (**Table II**) would be met. There is an interesting corollary to this scenario. One of the criticisms that is commonly leveled against alien UFO advocates is the lack of any hard evidence. "Hard evidence" could be defined, as something for which there is no other possible explanation. Carl Sagan was fond of saying, "Extraordinary claims require extraordinary evidence," and so far there really isn't any hard, extraordinary evidence. However this may be one of those rare instances in which the lack of evidence *is* evidence. It may be no hard evidence exists because the alien craft are only virtual. There would have to be at least one physical aspect to such a virtual craft and that would be the beam itself with its possible ring enclosure. A beam so configured could leave behind evidence of its presence such as burn marks or wilted vegetation.[20] Even so, it would probably still not be considered hard evidence.

One of the strongest driving forces behind our own astronomical research is the desire to know if others are "out there" like us, or even to know whether life actually exists elsewhere. It seems logical to assume that any alien culture would also be curious on this point. Any such culture, possibly hundreds, thousands, or millions of years older than ours would most likely have been working on this question for quite some time—possibly using technology like that envisioned above. By now they would have developed a list of star systems and associated planets on which they believe interesting developments are taking place, and maybe they have reached the point where they feel they need more than just a virtual presence.

Teleportation

As previously noted, interstellar space travel for biotic life involves logistical problems and safety hazards which, in spite of the many science fiction scenarios, we currently find to be insurmountable—at least as it is done on *Star Trek*. Could teleportation solve these problems? Teleportation, as we currently understand it, requires some kind of receiving device. This makes it fundamentally different from the remote sensing speculation described previously, although both rely

---

[20] [13] Figure 6

on the phenomenon of quantum entanglement.

Dr. David Darling, in his book, *Teleportation: The Impossible Leap*[05], presents a scenario originally conceived by physicist John Cramer of the University of Washington:

"... Cramer conjured up a scenario in which interstellar travelers take advantage of the technique in a vastly evolved form. Cramer imagines a spacecraft that sets out for the neighboring star Tau Ceti, 11.9 light-years away, carrying the macroscopic entangled states of similar states held in storage on Earth. Upon arrival after perhaps centuries or millennia of travel, robots unload the ship and set up a teleport receiver unit for each entangled state. When all is ready, the colonists step into the transmitter units where they join the stay-at-home entangled states and are destructively measured by the transmitter. The results of the measurements are recorded and sent by radio or light beam to the receivers at Tau Ceti. Twelve years later, Earth time, the beamed information is received at Tau Ceti, the transformations on the entangled states are performed, and the colonists emerge from the receiver units.

etching   Only Information Makes The Trip   R. L. Mason

For the teleported colonists, no subjective time at all has passed, and they have had a perfectly safe trip because it was known that the receiver apparatus had arrived and was fully checked out before the transmission began. The explorers can also return to Earth, with the same twelve-year gap in subjective time, using other entangled subsystems that were brought along for the trip to teleport in the opposite direction. As Cramer points out, this form of weird quantum jumping might not have quite the visceral appeal of warp-factor-nine starships, but it might be a much safer and more efficient way of exploring the stars."[21]

From the above quote one can see how a technically advanced alien civilization, with time on its hands (hands?), might "hopscotch" around the galaxy keeping track of interesting developments. However, before *you* sign up for this kind of adventure, there are some things you need to know:

First, you will need a quantum computer for this kind of leap because every particle of your body must be scanned, recorded, and stored in memory. This is far beyond the capability of any computer existing today. You will probably need such a computer at the receiving end as well.

Second, you should be aware that the transmitting process of quantum teleportation destroys the original you. *Only information makes the trip* and the receiver creates an exact copy of you at the other end.

---

[21] [05] p. 226

etching   R. L. Mason

Thirdly, another kind of teleportation does not destroy the original. You will still need your quantum computers, but this "classical teleportation" does not involve quantum entanglement, and can only promise to make a good replica of you, not an exact copy. But then there will, in essence, be two of you, the original and the replica. It doesn't take much contemplation to realize that this could lead to complications. It's probably a good idea to restrict the use of classical teleportation to inanimate objects. Restricted to that kind of use, it essentially becomes a replicator, which, if you will remember from *Star Trek,* can be a very handy thing indeed. One of these should be sent along to your destination with the original robotic mission that delivers the receiver. That way, when you arrive, it will be there to help you set up housekeeping.

It should be noted that quantum computers, quantum teleportation, and replication are not impossible — just technically very difficult. We have, in fact, already accomplished these things on a rudimentary level. The remote sensing (virtual spaceship) speculation may be somewhat more problematic.

The Moon, The Moon, The Moon

As we currently understand it, quantum entanglement is a fairly fragile connection. Interference by other particles or various forms of radiation can cause what is referred to as "decoherence." This can cause a degradation of the connection. A certain amount of degradation may be acceptable if all you are after is an image, as in remote sensing. This would be roughly analogous to experiencing static on a radio receiver or "snow" on a TV. Some kind of computer enhancement or correction could compensate for such a problem. However, if you were transmitting the essence of a biotic being, you would probably want interference to be minimal. Consequently, any alien culture that wished to establish a physical presence in our region of the Galaxy would probably try to avoid such phenomena as Earth's magnetic field, atmosphere, and Van Allen radiation belts. The Moon has none of these problems. In addition, the alien may wish to remain incognito to the maximum degree possible, and the Moon offers obvious advantages in that regard as well. They could use the far side of the moon, which we never see directly from Earth to mask much of their comings and goings. They may also find some of the deep craters at the poles, where we suspect water ice may exist, to be attractive. Maybe, for technical reasons, they would want to be directly opposite Earth. With adequate camouflage this should not be a problem. It would seem, however, that there is compelling logic for a subsurface facility. Especially since there is no atmosphere, magnetic field, or radiation belt to protect them from incoming meteors, high-energy particles, and cosmic rays. Stanley Kubrick did a nice job of envisioning such a facility in the movie *2001: A Space Odyssey*. Once their "colonists" arrived they could replicate their remote sensing equipment and use it with much greater accuracy than they could from some distant star system. A particle beam originating on the Moon could be at Earth in a matter of minutes, or even seconds. They could, of course, also replicate actual physical spacecraft for investigations on location and in person. When you think about it, it is hard to imagine more perfect circumstances for alien observation of developments on Earth than would be afforded by an outpost on the Moon.

Could we detect such an outpost? A subsurface facility on the Moon could possibly be detected from a low orbit using ground-penetrating radar, or by measuring magnetic and gravitational anomalies. We might also look for various types of radiation leaking from such a facility. In addition, events called Transient Lunar Phenomena (TLP) may be out-gassings of some kind[22]. Their existence has been known for a long time, but they are only recently receiving serious attention.

One of the good things about the absence of an atmosphere is it makes quite a low orbit possible. Even if this turned out to be a wild goose chase, it still might be worth doing from a scientific standpoint. In fact, some of this has already been done by NASA's Clementine mission (1994), the Lunar Prospector mission (1998-99) and the European Space Agency's SMART-1 (2005-6). If these missions *did* discover an alien base on the Moon it has yet to become public knowledge.

This is the appropriate point to mention some material I consider questionable, but which may contain elements of truth. Any number of websites deal with the subject of UFOs. One that appears to be particularly well organized is entitled: UFO Evidence (ufoevidence.org) and deals with everything from A to Z connected with UFOs. If you click on *Astronaut Sightings*, some of the items that pop up purport to be transcripts of Apollo Astronaut conversations while on the Moon.[23] These excerpts apparently pertain to encounters with hard-to-explain structures and sightings of unidentified

---

[22] [18] p. 358

[23] http//www.v-j-enterprises.com/astro2.html
Also: *Celestial Raise* by Richard Watson and ASSK, 1987 p. 147

craft. The most startling of these quotes are exclamations by Armstrong and Aldrin of Apollo 11 concerning unidentified spacecraft. The communication of these exclamations was interrupted for approximately two minutes which NASA attributes to the overheating of a TV camera.

However, according to this website, a large number of ham radio operators were tuned in directly and heard what was said after NASA "pulled the plug." These include the following and much more:

"I say that there are other spaceships!"

"They landed here!"

"They're lined up in the other side of the crater!"

"There they are and they're watching us!"

I know this sounds too incredible to be true . . . and *maybe* it is.

Chapter 6
But Wait A Minute. . . Humanoids?

If we proceed as if the Alien Speculation in the previous chapter is true, whether physical or virtual, then questions naturally arise about the operators of these craft. The craft could, of course, be fully automated, but many of the reports that deal with this subject describe autonomous beings.

Dr. Hynek gave Close Encounters of the Third Kind (**Table I**) his highest strangeness rating; in fact, he expressed some reluctance to even tackle the subject:

> "We come now to the most bizarre and seemingly incredible aspect of the entire UFO phenomenon. To be frank, I would gladly omit this part if I could without offense to scientific integrity: Close Encounters of the Third Kind, those in which the presence of animated creatures is reported, (I say 'animated' rather than 'animate' to keep open the possibility of robots or something other than 'flesh and blood.') These creatures have been variously termed 'occupants,' 'humanoids,' 'UFOnauts,' and even 'UFOsapiens.'"[24]

His reluctance is understandable. When I created **Table II: Commonly Mentioned Characteristics**, I left out descriptions of aliens because it didn't seem to fit with the other items, and, therefore, needed separate treatment.

---

[24] [13] p. 158

"Humanoids" seems to be the term in vogue at this writing and the descriptions *are* fairly consistent. The difficulty that anyone with a rudimentary knowledge of evolution has with these reports lies with the improbability that beings from a distant star system with an entirely separate biological history would look anything like us. It smacks of human conceit. *If other intelligent, technologically advanced beings exist, then they must look something like us.* It's the same kind of thinking, differing only by degree, that caused Michelangelo to represent God on the ceiling of the Sistine Chapel as a human being. Stephen Jay Gould in his book, *Wonderful Life*[11],[25] points out that we are the result of a very long string of hit-or-miss events, low probability occurrences, fortuitous timing, and happy accidents; and even here on Earth, if the clock of life were reset to the beginning and rerun, the probability that human beings, as they exist today, would emerge again at the end of some three and a half billion years of evolution is as close to zero as you can possibly get. The counter argument is the one of *convergence* (like conditions produce like results), but to me this still seems inadequate to explain the striking resemblance that most descriptions of alien humanoids bear to us.

In addition, Dr. Hynek continues:

> "Another thing bothers us: the humanoids seem to be able to breathe our air and to adapt to our air pressure and gravity with little difficulty. Something seems terribly wrong about that. This would imply that they must be from a place — another planet? — very much

---

[25] [11] p. 320

like our own. Perhaps our own? But how? Or are they robots, not needing to adapt to our environment?"[26]

He lists only a small number of prototype reports in his book, but they *are* quite compelling and, as mentioned previously, the descriptions of humanoids *are* fairly consistent. So, in light of the above, how do we account for such reports?

Humanoid Speculation #1: Myth

These descriptions originally spring from science fiction or someone's hallucination. They were imitated by others either consciously or unconsciously, and were eventually picked up by Hollywood for inclusion in such films as *Close Encounters of the Third Kind* by Steven Spielberg. There, the image was refined and distributed for mass consumption. Today the humanoid image has become a modern mythology and this mythology is tapped whenever alien contact is reported for whatever reason.

Humanoid Speculation #2: Us

The aliens resemble us because they *are* us but in a highly evolved future state. They are, in fact, time travelers returning to *this* time to do research on their distant past. Today we haven't the slightest idea how this could be done. With the possible exception of cosmic wormholes, a far out concept which I'll not go into here, I am unaware of anything in currently established science that even points in this direction. We *are* aware that time can proceed at different rates for

---

[26] [13] p. 159

bodies traveling at different fractions of the speed of light, but that is moving into the future not the past. Even if the aliens did know how to travel into the past, it would probably be a dangerous game. Any tinkering with an event of the distant past is going to be the cause of other events, which in turn, will be the cause of still more events, and what results is an expanding "cone" of cause and effect change proceeding into the future. The alien who makes some, even slight, change to the distant past runs the risk of suddenly finding circumstances in his *own* time which are not familiar to him, and in the worst case scenario he could even eliminate his own existence.

Humanoid Speculation #3: Designer Beings

The aliens are designer beings especially created to resemble us in case chance encounters occur during their observation process. Maybe they are "born" from replicaters at an outpost on the Moon. They might be some amalgam of biotic, electronic, mechanical, and chemical construction. They may be robotic "worker bees" for the real alien who actually looks more like an octopus.

Chapter 7
Extra Dimensions & Hidden Worlds

The field of quantum physics (mechanics), briefly mentioned in Chapter 5, is a difficult subject to wrap your mind around. In fact, it may be impossible. No less a luminary than Richard Feynman, a Nobel Laureate in this field, is often quoted as saying, "I think I can safely say that nobody understands quantum mechanics."[27] Things just don't behave on the level of the very small in the same way they do in the macroscopic world, or so it seems.

Much of what happens on the level of the very small is counter intuitive to us. Things wink into and out of existence at subatomic dimensions, and the very nature of a particle seems to depend on whether or not it is being observed. Quantum entanglement, as described in Chapter 5, is one of the strangest aspects of quantum mechanics. How could two widely separated (but entangled) particles communicate instantaneously? We must be missing some very basic ingredient, *another dimension perhaps?*

Thought Experiment #2

Suppose an irregularly shaped, three-dimensional object were to pass through a two dimensional world (see Figure #1). Beings living in the 2D world would see only that portion of the object in their world. In 2D, the object may manifest itself

---

[27] Feynman, Richard Phillips *The Character of Physical Law.* MIT Press, 1967, p. 129

as more than one object, however in 3D it is all one object. Now add on another dimension. Suppose a 4D object were to pass through our 3D world. Could this explain entanglement? In other words, could two entangled particles appear to communicate instantaneously in 3D because they are actually connected in 4D?

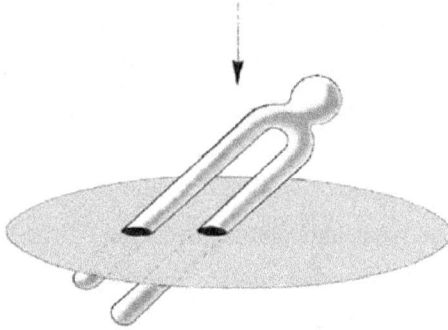

(Figure #1).
A 3D object passes through a 2D world

Knowledge and use of a fourth, spatial dimension would certainly be a handy thing for a UFO of alien origin. It could go a long way toward explaining items number 8, 9, and especially item number 10, of **Table II**.

A current "Holy Grail" of physics is to be able make the laws that govern the Universe on a large scale (general relativity, etc.) jibe with laws that govern the Universe on a small scale (quantum mechanics). A single theory that would accomplish this task is sometimes referred to as a Theory of Everything (TOE). Some of the best minds in science have been working on this problem for almost a century now without any obvious solution.

At this writing, the most popular candidates for solving the problem are *superstring* theory and its cousin *M-brane* theory. Both of these show some promise as mathematical constructs, but they require the existence of six or seven additional spatial dimensions. After finishing Brian Greene's book *The Elegant Universe*[12] that deals with these theories at length, my initial reaction was, if ever there was a violation of Occam's Razor this is it. Occam's Razor is the scientific maxim originally stated by William Occam (1285 -1349)

> "It is vain to do with more what can be done with less" and *"Entities should not be multiplied beyond necessity"*[28] [emphasis added]

It has been rephrased and paraphrased over the years. Some of the forms it has taken include:

> "If a simple explanation will do, it is idle to seek a complex one"

and,

> "When given a choice between several explanations for a given phenomenon the simplest and most straight forward is the one most likely to be true."

The problem is superstring theory and M-brane theory are almost the only games in town. One wonders, though, if they don't leave the realm of science behind. They appear to be more metaphysical than physical.

---

[28] [20] p. 162

What is being proposed is perilously close to a synthetic *a priori* (see Appendix A) which would put it in the company of religious entities. No empirical observation can be made, and science is, after all, empirical. There *is* one other theory called "Loop Quantum Gravity"[29] but at this point (2007) it hasn't attracted anywhere near as much attention as the first two.

Similar problems exist for current cosmological theories involving so-called "dark matter" and "dark energy," but at least in the case of these two, they originate as the result of anomalous observations of the physical world. In addition, it is only fair to point out that these two are only speculations or conjectures as opposed to superstrings and M-branes, which are highly evolved theories.

In the case of dark matter, there *is* an alternative that may be simpler. In its original form it is known by the acronym MOND for Modified Newtonian Dynamics.[30] In a later incarnation it is called TeVeS for Tensor, Vector, and Scalar. The problem for *these* theories is, for them to be right, both Isaac Newton and Albert Einstein have to have been . . . well . . . not quite right, and that is going to be a tough sell.

Other Universes?

As mentioned above, scientists have long been frustrated by their inability to bring together general relativity and quantum mechanics into one unified theory. But why do they feel so strongly that this is necessary? Is there some basic reason why these two can't stand separately yet still work together like symbiosis? Perhaps the desire to find a unifying

---

[29] [24] p. 250

[30] Frank, Adam "Gravity's Gadfly" *Discover*, August 2006 p. 33

TOE is more indicative of human nature than the nature of the Universe. We may be predisposed to see things in this way. Jana Levin, in her book, *How the Universe Got Its Spots*[17], speculates on this idea from a slightly different angle:

> "But I can't help but wonder if there isn't a much more radical and deeper role for chaos in theoretical physics. Maybe there are no symmetries, no firm laws, no rigorous order. Maybe our experience of order and the laws of physics is the order that precipitates from complexity. Maybe there isn't an ultimate law, one fundamental symmetry, but instead many, a proliferation of possible laws, and the seeming symmetries that guide our perception of the forces emerge from the collusion of a democracy of quantum theories."[31]

Why couldn't the incompatibility of the two basic theories be considered an empirical observation that is trying to tell us something? M-brane theory leads us to the possibility of parallel universes *each with its own peculiar set of physical laws,* but requires the existence of additional dimensions to do this. But what if the separation of universes is not a matter of extra dimensions but instead a matter of scale. Perhaps the sub-atomic scale where quantum mechanics rules should be considered a different universe, a "microverse" so to speak, complete with its own set of physical laws. This, of course, leaves open the possibility of a universe on a much larger scale than ours as well, a "macroverse" with it's own set of governing laws. Could dark energy and dark matter be

---

[31] [17] p. 187

manifestations of such a macroverse?

I am reminded of the ditty by Jonathan Swift that my father used to quote:

"Big bugs have little bugs upon their backs to bite them.
Little bugs have littler bugs and so, ad infinitum."

Life on Lilliputs

It seems apparent that most bodies of any size in the Universe, from asteroids to black holes, formed in about the same way — by gravitational accretion. Some asteroids and maybe some small moons came into existence as the result of collisions between larger bodies, but their origin still goes back to some dust cloud nursery where a little seed became a "snowball" and then went on to bigger things. The dust clouds themselves are the result of explosions such as supernovae and are the leftovers from a previous generation of stars. Depending on the density of the dust cloud, and hence the amount of material available at a particular locale, different sized bodies emerged. Thus we would expect that a complete distribution of sizes should be present in any galaxy. Galaxies themselves are probably examples of gravitational accretion on a grand scale, and many are thought to have a massive black hole at their center that is gobbling up stars.

In fact, we *do* see a complete distribution of sizes when we look out at the Universe. The larger bodies, upon reaching a certain threshold in mass, "turn on" by a fusion reaction at their core. We see these bodies because they emit electromagnetic radiation in the visible portion of the spectrum (light) and we call them stars. Smaller bodies that never reach the threshold for "turning on" may be found in orbit around the stars and these bodies are seen (at least in the solar system) by the light they reflect. We call these bodies planets, moons, asteroids, etc. It is quite possible, however, that many smaller bodies form in locations between stars and that causes them to be independent of the influence of any star. We don't see them because they neither emit nor reflect light. What do you call such a body? "Star" doesn't seem right, nor does "planet."

Astronomer Harlow Shapley, in a piece entitled "Life on Stellar Surfaces" that was included in the book *The Scientist Speculates*[10][32] calls these bodies Lilliputian stars or Lilliputs, borrowing the term from *Gulliver's Travels*. Writing in 1962, he suggests that the numbers of Lilliputs may be quite large, as much as ten or even a hundred times the number of visible stars. If a body has less than one tenth of the mass of our Sun it will fail to turn on by nuclear fusion. Such bodies are very difficult to find, but thanks to recent advances in infrared astronomy some have been observed. They "shine" in infrared from internal heat left over from their original formation, gravitational compression, and low-grade fission reactions that result from the presence of elements such as uranium, thorium, radium, and potassium-40. The more recent term "brown dwarf" has been used to describe these Lilliputs and appears to apply to those bodies less than about seventy-five times the mass of the planet Jupiter.

If planet Earth were a Lilliput its oceans would be frozen solid. Earth depends on radiation from the Sun to keep them in a liquid state. Earth does have some internal heat, of course, but not enough to keep its surface water liquid in the absence

---

[32] [10] p. 225

of the Sun. Larger bodies, however, somewhere between the size of Jupiter and the upper limit for a brown dwarf, may have sufficient internal heat to keep surface water liquid in the absence of any nearby star. Such a world would be dark in the visible portion of the spectrum, and it would (unlike Earth) retain all the lighter gases of its primordial atmosphere. Why couldn't life evolve under such conditions?

In 1969 the American Association for the Advancement of Science held a symposium on the subject of UFOs. It was organized and hosted by the astronomers Carl Sagan and Thornton Page. Papers presented at this symposium were later gathered together in a book edited by the hosts and entitled *UFO's: A Scientific Debate*[12]. One of the papers presented, and included in the book, was a piece by astronomer Franklin Roach that expanded upon the previous work of Harlow Shapley as outlined above.[33] Roach, by interpolating between the smallest and the largest bodies known, was able to arrive at an estimate for the number of Lilliputs in a range of sizes. Based on the mean density of such objects in the Galaxy, he determined that the nearest Jupiter-sized object could be somewhat less than one light year from the Sun. Since Shapley had estimated that a Lilliput would have to be slightly larger than Jupiter to have the required amount of internal heat to spawn life, Roach suggested that the nearest life outside our Solar system might be found on a Lilliput rather than a planet in orbit around another star.

Carl Sagan, in a critique of the above speculation,[34] claimed that the temperature gradient between the heat *source*

---

[33] [12] p. 23
[34] [12] p. 291

and the heat *sink* would be insufficiently steep to be useful as an energy source for potential life (this same critique had been leveled against Shapley's speculations some seven years earlier). However, that was 1969, in 1978 the first undersea hydrothermal vent was discovered near the Galapagos Islands, and there — big as life — *was* life, and it was abundant in the complete absence of Sunlight. Since that time a substantial number of hydrothermal vents have been discovered and each one has it's own ecosystem together with a cast of living characters. In fact, it is now widely hypothesized that life on Earth could have originated at vents like these. I see no reason, therefore, that life couldn't originate on a Lilliput, in the same way especially if it had all the right prerequisites.

I am a great admirer of Carl Sagan, but I think there is a good chance he got it wrong that time. This is no reflection on Sagan, however, it's simply the way science works (see Appendix A).

Chapter 8
A Close Encounter of the Third Kind

On June 26 and 27 in 1959 the Rev. William Booth Gill observed a UFO in Papua, New Guinea at the Boianai Mission of which he was the resident Anglican missionary. This is a well-known case[35] but it contains elements that could bear on various speculations in this book.

On the day prior to the sighting Father Gill had written a letter to the Rev. David Durie that reads as follows:

"Dear David,

Have a look at this extraordinary data. I am almost convinced about the 'visitation' theory. There have been quite a number of reports over the months, from reliable witnesses. The peculiar thing about these most recent reports is that the UFOs seem to be stationary at Boianai or to travel from Bioanai. The Mount Pudi vicinity seems to be the hovering area. I myself saw a stationary white light twice on the same night on 9 April, but in a different place each time.

I believe your students have also sighted one over Boianai. The Assistant District Officer, Bob Smith and Mr. Glover have all seen it, or similar ones on different occasions again, over Boianai, although I think the Baniara people said they watched it travel

---
[35] [13] p. 168 & [14] p. 206

across the sky from our direction. I should think that this is the first time that the 'saucer' has been identified as such.

I do not doubt the existence of these 'things' (indeed I cannot, now that I have seen one for myself) but my simple mind still requires scientific evidence before I can accept the from outer space theory. I am inclined to believe that probably many UFOs are more likely some form of electric phenomena, or perhaps something brought about by the atom bomb explosion, etc.

That Stephen should actually make out a saucer could be the work of the unconscious mind as it is very likely that at some time he has seen illustrations of some kind in a magazine, or it is very possible that saucers do exist, but it is only a 50/50 chance that they are not earth made, still less that they should carry men (more likely radio controlled), and it is still unproven that they are solids.

It is all too difficult to understand for me; I prefer to wait for some bright boy to catch one to be exhibited in Martin Square. Please return this report as I have no copy and I want Nor, (Rev. Norman Crutwell) to have it.

Yours,
Doubting William
Anglican Mission, Boianai.

On the very next day Doubting William was converted as indicated from the following note:

"Dear David,

Life is strange, isn't it? Yesterday I wrote you a letter, (which I still intend sending you) expressing opinions re: The UFOs. Now, less than twenty-four hours later I have changed my views somewhat. Last night we at Boianai experienced about four hours of UFO activity, and there is no doubt whatsoever that they are handled by beings of some kind. At times it was absolutely breathtaking. Here is the report. Please pass it round, but great care must be taken as I have no other, and this, like the one I made out re: Stephen, will be sent to Nor. I would appreciate it if you could send the lot back as soon as poss.

Cheers,
Convinced Bill"

The report produced by Rev. Gill runs to eleven pages single spaced and is signed by 29 of a reported total of 38 witnesses including teachers and staff at the mission. It covers similar sightings on two successive evenings (June 26 & 27, 1959) for a total of about five hours beginning at just after sunset. It describes a brightly lit object as big as five full moons placed side by side. The object was sighted over the water at an estimated elevation of 150 meters and at an angle of about 30 degrees above the horizon from the position of the viewers. It appeared to be a solid circular object with a wide base and a narrower upper "deck" and it had four legs that protruded from below. *On a number of occasions a shaft of blue light could be seen projecting up into sky at an angle of about 45 degrees.*

From time to time as many as four "men"could be seen moving about on the deck. They looked to be human beings but the distance was probably too great to be sure. They appeared to be occupied with some apparatus at the center of the craft and they could be seen to bend over and their arms moved up and down as if adjusting "or setting up"something. Rev. Gill's notes include this interesting passage:

> "One figure seemed to be standing looking down at us (a group of about a dozen). I stretched my arm above my head and waved. To our surprise the figure did the same. Ananias waved both arms over his head then the two outside figures did the same. Ananias and self began waving our arms and all four now seemed to wave back. There seemed to be no doubt that our movements were answered. All mission boys made audible gasps (of either joy or surprise, perhaps both)."

Father Gill's notes also include a sketch, a tracing of which is shown in Figure #2 below. The sketch was retrieved from the website entitled: *UFOs at Close Sight* (ufologie.net). It can also be seen along with an artist's impression of the encounter at *UFO Evidence* (ufoevidence.org).

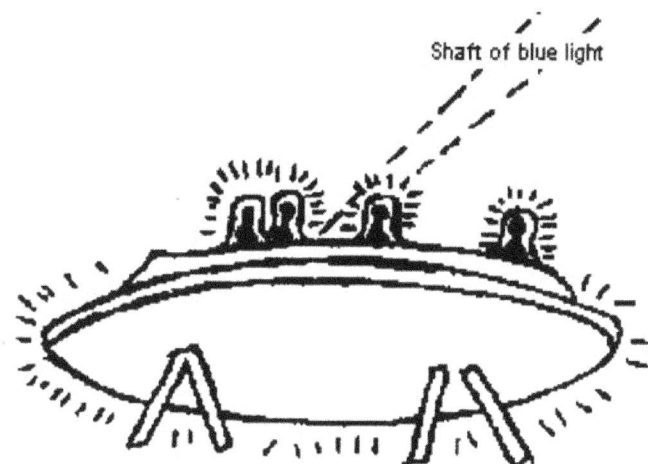

(Figure #2)[36]
Tracing from the original sketch by Father Gill

Much more detailed accounts of this case can be found at these two web sites as well as in Dr. Hynek's book *The UFO Experience: A Scientific Inquiry* [13].

This case manages to be, at once, both one of the strangest, and one of the best documented of the so-called classic sightings. It gains its credibility from the following:

1. The existence of the before and after letters quoted above.

2. The character and demeanor of Father Gill, a highly respected person on the Island and someone whom all investigators, including Hynek, found to be very impressive.

---

3.   The detailed report written by Father Gill at the time.

4.   The number of witnesses.

5.   The duration of the sighting and its context within a wave of sightings in that general time period and location.

6.   The extent of the follow-up investigations. Hynek and his staff, in particular, went to great lengths including visits to the sight and interviews conducted over a long time period.

It is somewhat disappointing and surprising given the number of witnesses and the duration of the sighting that no photographs were made. Surely there must have been a camera available somewhere in the mission. Even so, it is a very intriguing case. The appearance of humans or humanoids would seem to rule out a natural explanation.

Father Gill Case: Domestic Speculations

One suggested explanation is a mirage of a brightly lit squid boat somewhere offshore complete with a false horizon. This would seem to have a degree of merit except that Rev. Gill's report and his sketch all indicate that the object was seen from below and it had legs protruding beneath it.

Rev. Gill himself suggested another domestic explanation:

"I realized, of course, that some people might think of this as a flying saucer, but I took it to be some kind of hovercraft the Americans or even the Australians had built. The figures inside looked perfectly human. In fact, I thought they were human, that if we got them to land we would find the pilots to be ordinary earthmen in military uniforms and we would have dinner with them."

Father Gill Case: Alien Speculations

I am particularly struck by the description of a "shaft of blue light that shown at an angle up into the sky. It is also shown in the sketch. Could this be a particle beam organized as I have speculated in Chapter 5? Could this be the elusive virtual spaceship?

Maybe a unique combination of atmospheric conditions and lighting circumstances allowed the observers in this case to see something that is not usually visible. Perhaps, what was observed was a hologram of the storage ring in some alien laboratory off planet Earth somewhere. The alien lab sent a particle beam organized in the manner previously speculated for the purpose of remote sensing. The beam attenuates as it encounters Earth's atmosphere and terminates in a bright plasma display as it eventually collides with molecules of air. Superimposed on the radiated light is the synchrotron emission of the particles (electrons?) that are in orbit around, and sheathe the carrier beam. This emission actively senses the targeted environment, and the orbiting particles are entangled with those in the alien laboratory storage ring. Maybe the storage ring is housed in some kind of chamber that allows the synchrotron radiation that *it* emits to produce a holographic image (via entanglement) of what is sensed of the environment at the other end. Thus for a brief period of time a two-way communication link was established between Father Gill, his group, and four technicians in an alien lab operating a remote sensing device (virtual spaceship?) somewhere off Earth.

This speculation, though complex, seems plausible except I am still bothered by the description of the aliens appearing as human beings. The probability of that seems very small to me. The only explanation I can come up with is the *designer beings* that I suggested almost in jest under **Humanoid Speculation #3** (page 56). If these beings *are* being designed from scratch it would be a good idea to give them a high tolerance for radiation, because they are going to encounter quite a bit walking around on that synchrotron storage ring.

Father Gill Case: A Natural Speculation After All

Quite some time after I finished writing the above I was lying awake in the middle of the night unable to go to sleep and my mind wandered back to this puzzling case. The more I thought about the mirage of a brightly lit squid boat the more I became interested in its possible veracity. Then I had an idea. We all have read accounts of mirages in a desert setting where witnesses tell of seeing the skyline of a city hundreds of miles away above a false horizon and just below this false horizon is the same skyline but upside down. So we know that a mirage is capable of inverting an image. The Father Gill case, however, is over water not a desert. If the squid boat were working in a flat calm, sometimes called a "glass-off" by sailors, then it would be reflected in the mirror-like surface of the water. Looking at this reflection directly you would see the whole scene inverted. Now, if this reflection were viewed from a great distance courtesy of a mirage that had a false horizon running through it then part of the image, which is already inverted by reflection, will be inverted again by the mirage. I think this may be why it appeared to Father Gill that he was looking up at figures that were seen from the waist up only. What he was seeing was the reflection of the figures re-inverted to look right side up.

Parts of the image were inverted only once, for instance the spars of the boat's fishing rig were inverted by the reflection to appear as landing gear of a spacecraft or hovercraft. The rest of the image on the other side of the false horizon was inverted twice to appear in an upright configuration. The shaft of blue light, which appeared to shine up into the sky, was actually a bright light used to illuminate the working area of the deck and to attract the squid. It was probably a carbon-arc lamp that was occasionally reflected off the water as the boat rolled slowly in the slight undulation of the surface.

In short what was seen, and accurately reported, by Father Gill was a mirage of a reflection, and what was seen from the boat was a reflection of a mirage. This has got to be an extremely rare circumstance and it's no wonder it was probably misinterpreted.

Chapter 9
Where Are They, and Why?

"Where is everybody?" This has become known as Fermi's Paradox and is based on a remark made by Enrico Fermi to his colleagues over lunch at Los Alamos in 1950. Based on some idle conversation about flying saucers (a hot topic in those days) Fermi apparently went through a rapid and complex mental calculation of the kind for which he was famous. In this case it was a compound probability calculation similar to Drake's Equation. His "order of magnitude" conclusion so startled him that he blurted out his question right in the middle of lunch where conversation had since passed on to other subjects. Fermi had concluded not only that other extraterrestrial civilizations should exist, but they also should be here by now.

As previously mentioned, any alien beings capable of making interstellar voyages to visit Earth are bound to be technically advanced when compared to us. Such an advanced — and probably much older — civilization should have no problem making itself known in an explicitly overt and unambiguous way if it so desired. However, this has definitely *not* been the case. So we are forced to conclude one of the following:

1. They do not exist; or

2. They exist but are not here; or

3.   They exist and are here but only in a virtual sense (Chapt. 5); or

4.   They exist and are here but wish to remain covert; or

5.   Some combination of (3) and (4).

There is, of course, the distinct possibility that they don't exist and this is, perhaps, the simplest solution to Fermi's Paradox, but if we accept *that*, it's the end of speculation. If we decide that they exist but are not here then it's back to our SETI endeavors and that's the end of speculation about presence. So let's not go to those places. This book is, after all, at least partly about speculation. If we decide that they *are* here, either virtually or physically, then why do they wish to remain covert? Ah! Plenty of speculation is possible on this point:

Motive Speculation #1: Scientific Investigation

They find us interesting from a scientific point of view only and have no interest in direct contact with such a primitive species for commercial or any other reason.

Imagine that you own a terrarium. It's a good-sized tabletop model, which you placed next to a window for sunlight. It has a supply of water, and a variety of plantings along with an assortment of "wee beasties" including a colony of ants. You are an entomologist and you study the behavior of the ant "civilization." You watch as they forage for food, and you can see them dig their tunnels through the glass of the terrarium. You note how they feed the queen and take the eggs that she lays to special incubation chambers designed for that purpose. You find this all very interesting and you marvel how their social structure has enabled their survival for millions of years. *But,* would you wish to make *your* presence known to them by attempting communication? If you did the ants would probably panic, run about helter-skelter, make pathetic attempts at self-defense, try to find a safer place for themselves and the eggs, etc. Why would you wish to disrupt their normal life routines to such a degree? Besides, what would an ant, with its extremely limited view of reality, have to say that would be of interest to you?

Motive Speculation #2: Monitors

We are candidates for inclusion in a galaxy wide treaty organization and they are keeping track of our progress as an emerging civilization. They haven't contacted us yet because they don't think we are ready. I think they would be right on that point. Any beings that can't even get along with themselves on their own world aren't going to be great signatory to a treaty of such grand scale.

Maybe they are helping the process along with a very long range plan to gradually get us used to the concept of other intelligent beings and civilizations in the Universe, and these helpful hints are what is being observed.

Motive Speculation #3: Long Range Planning

Most stars have a habitable zone where a planet would have water predominately in a liquid state. Earth sits smack dab in the middle of the Sun's habitable zone. Some astronomers believe galaxies have habitable zones as well[37] and an estimate

---

[37] [27] p. 27

has recently be made that the average age of Sun-like stars in the habitable zone of the Milky Way is approximately one billion years older than the Sun.[38] The average life span for Sun-like stars (medium sized stars) is about 10 billion years. Our Sun, at about 4.6 billion years is, therefore, a little less than half way through its life cycle, fairly young, so to speak. From the forgoing, however, it should be obvious that there exists the possibility of other civilizations whose stars are much nearer the end of their life. We are aware of *our* situation and, of course, they would know *theirs*. Perhaps they still have a billion or two years left. That's still a lot of time, but maybe they have begun to think about their situation whereas we have not.

We here on Earth have only recently become aware that dinosaurs, which dominated life here on Earth for millions of years, were almost totally wiped out (birds remain) by an asteroid impact some 65 million years ago. This was a wakeup call and we *have* begun to do something about it. We have started tracking and cataloging near Earth asteroids so that, at least we might have some advanced warning. Whether we could do anything about it is another question. A truly advanced civilization whose star was nearing the end of its life, might be even better at long range planning than we are and could be expected to be looking around for a new home. Perhaps, they are restrained by the grand intragalactic treaty from just coming here and taking Earth as they probably would be capable of doing. Perhaps, there is something akin to the *prime directive* that was often mentioned in episodes of

---

[38] Lineweaver, Fenner, & Gibson "The galactic habitable zone and the age distribution of complex life in the Milky Way" *Science* , 303, p. 59. 2004

Star Trek. Possibly, there is a galactic history of developing technological civilizations having short life spans, especially when their sociological skills lag far behind their technical skills. So they have come here and established an outpost where they can be ready to move quickly should we do ourselves in, or should some natural disaster overtake us. At any rate, such an outpost could have been established a very long time ago and they are poised and ready to take over our beautiful little blue planet the moment *we* disappear.

Where are they physically? If they wished to remain covert then they would locate someplace where we are not. I have already discussed the advantages of the Moon, but maybe they would find Mars to be more hospitable. Mercury and Venus would seem less likely. Under the seas is, I suppose, a possibility also. In addition, it has been suggested that a large spaceship parked in the asteroid belt between Mars and Jupiter would be almost impossible to detect, and one school of thought, championed by some Russian scientists, suggests that the moons of Mars might be artificial! One other place might be a candidate, but more on that in Chapter 13.

Motive Speculation #4: We Are Entertainment

In his account in Chapter 2, Gordon Chism got the strong impression that the UFO he encountered had a sense of humor. Others have mentioned this as well. Pilots of military and commercial aircraft often state they felt like they were being toyed with in a kind of benign cat and mouse fashion.

My partner, Evie Wilson, tells a story that dates back to the 1960s when she was married, raising small children, and living in Los Angeles. Another family with whom they were friendly used to visit them from Northern California. The husband worked as an engineer in Silicon Valley for a firm that was developing the laser in the early days of that technology. On one occasion he brought along a prototype, hand held model for show and tell. Everyone was duly impressed with the laser's capabilities and they amused themselves by hiding and shining the laser into the neighbor's house across the street, placing a dancing red dot on the neighbor's living room wall. Evie's family and friends found the antics of the neighbors to be very entertaining as they watched them scratch their heads and chase the dot.

Is there a parallel here? Evie's group was in possession of technology that was unknown to the neighbors, and they used this disparity to entertain themselves. This could very well be the case between an alien culture and us. I speculate in Chapter 8 that a synchrotron storage ring located in a special chamber, off Earth somewhere, might allow a holographic image to be created of the environment found at the target end of their particle beam. Could this also be a theater of some kind used for the entertainment of aliens? One is reminded of the *Holodeck* often seen in episodes of *Star Trek*. Perhaps the aliens use such a facility much as we use TV. It could receive live images as they happen, or it could access a library of prerecorded entertainment.

Maybe Earth is a galactic park much as we have national parks, or possibly a "wild animal" reservation with strict prohibitions against development. Viewing and visitation are only allowed under highly controlled conditions.[39]

---

[39] This is sometimes referred to as "The Zoo Hypothesis"

Motive Speculation #5: Communication Is Impossible

Perhaps, they are so radically different from us physically, culturally, and every other way that there is no hope of communication, and having come to this conclusion the aliens have decided to simply avoid direct contact. If this seems implausible I would point out that we are unable to communicate to any great extent with dolphins who are our fellow Earthlings, mammals, generally considered to be intelligent, and in possession of a language to boot. It will be noted that this speculation is only slightly different from Motive Speculation #1.

Chapter 10
Getting Beyond Anecdotal Evidence

It really is a shame that serious investigation into the UFO question carries with it such a stigma. It is understandable, but lamentable. Being retired helps, but working scientists and engineers whose professional reputations depend on the esteem of their colleagues avoid the subject like the plague. It takes a brave individual to tackle the problem, and that is another reason I admire Dr. J. Allen Hynek.

With sufficient professional interest and funds some things could be done to resolve the question. A network of automated observation posts, for instance, would be helpful. An example of this kind of thing does exist, but for a different reason. Generally referred to as fireball networks, they consist of video cameras trained on the sky full time and equipped with "fisheye" lenses or reflectors that image and record the entire sky from the position of their installation. A system of these installations spread over a geographic area with overlapping areas of observation is called a network and their purpose is to record the trajectory of large meteors (fireballs) as they enter and burn up in Earth's atmosphere. When two or more stations record such an event, specialized computer hardware and software can calculate the most likely area on Earth's surface where any remains may be found. The networks concentrate their efforts on large meteors or "bolides" because those are most likely to make it all the way to the surface thus becoming *meteorites*. Calculations can also be made that will indicate where in the sky the fireballs originated. Networks of this type now exist in the United States, Canada, Europe, Australia, and Japan. This has all come to pass in just the last half dozen years with the original installations dating from about 1998.

One such network with cameras in the U.S. and Canada is the Sandia Fireball Network[40] headquartered at the Sandia National Laboratories near Albuquerque, New Mexico. I sent repeated e-mails (see Appendix B.) to those in charge of this network asking if their network ever picked up "fireballs" that;

a.) change course during observation, or

b.) move too slowly to be a meteor, or

c.) stop and hover, or

d.) accelerate instead of decelerate, or

e.) exhibit any of the other characteristics commonly mentioned in association with UFOs.

I never received an answer despite the fact that their website states that most questions are answered within two working days. Most likely this is a result of the stigma mentioned at the beginning of this chapter along with my lack of affiliation with any group having standing within the scientific community. Much less likely is the possibility that I came too close to a classified objective of the network.

Another technology having come into existence during approximately the same period is the development of the robotic telescope.[41] It is no longer necessary to stay up all night

[40] Gamble, Jim "All-Sky Firevball Network" *Astronomy* May 2004, p. 76

[41] Polakis, Tom "Robotic Observing" *Astronomy* May 2004, p. 80

shivering in the cold while gazing through the eyepiece of your telescope. Telescopes fitted with digital cameras (CCDs) record and feed the desired image directly to a computer on a preprogrammed schedule while you are warm and sound asleep in your bed. This may take some of the romance out of astronomy, but it *is* much more efficient. The computer can also screen the images for you and display only those containing the type of phenomena you are seeking.

A marriage of these two developments seems a reasonable possibility and could greatly enhance the evidence researchers have to work with regarding UFOs. The all-sky networks could be programmed to recognize objects with movements typical of UFOs, namely most of the items included in **Table II**. When such an object is recognized, an alarm would be sent to robotic telescopes that would immediately slew to the positions indicated and record what is there at high magnification. An accumulation of this kind of data, especially if used in conjunction with radar and radiation detectors, would certainly be superior to anything currently in existence.

In 2004, the Mexican air force photographed some fast moving phenomena[42] that exhibited many of the movement characteristics listed in **Table II**. The images were made with a video camera equipped with an *infrared* lens. The phenomena were not visible to the naked eye. Only heat emanating from the source was recorded and no shape was discernible. It is worth noting that the photos were made in a region of Mexico that includes a lot of atmospheric gases due to the presence of oil and gas refineries.

---

[42] Castillo, E. Eduardo, Associated Press, "Mexican UFOs. . ." *The Press Democrat"* May 14,2004

Another kind of observatory with a full-time wide-angle view of the sky is just now (2006) coming on line in Argentina. Located on the pampas, the Pierre Auger Observatory[43] is designed to study cosmic rays and is a hybrid of two types of equipment. A network of water tanks each holding 3000 gallons is spaced about 1.5 kilometers apart, thus covering a very large area. The tanks are completely dark inside and contain photo multiplier tubes that record the light generated when a charged particle passes through the water. These operate full time. In addition, on occasions when the night sky is clear with no Moon, numerous high-resolution cameras record light in the ultraviolet portion of the spectrum. Ultraviolet light is emitted when high-energy charged particles interact with nitrogen in the atmosphere. Such light is not visible to the naked eye, but *can* be detected by these cameras.

So, it seems that our proposed automated UFO detection network should include this kind of capability as well as the infrared lens mentioned above. With sufficient interest and funds, putting such an observatory in orbit in the form of a specialized satellite would merit consideration.

A network with all the capabilities envisioned in this chapter could take us a long way down the path toward determining whether we are dealing with natural, domestic, or alien phenomena even though it may not pin the answer down precisely.

---

[43] http://www.auger.org/observatory/observatory.html

Chapter 11
Alien Particle Beam Recap

Dr. Hynek, writing in a foreword to the book *The UFO Controversy in America* [15] by David Michael Jacobs, summarizes UFO characteristics as follows:

"The reported ability to execute trajectories, often but not always silently, that no known man-made craft could generate or follow; the ability to hover, and then to accelerate to high speeds in the order of seconds (and generally without a sonic boom); on occasion to change shape, and to produce durable physical effects on both animate and inanimate matter. To be, on occasion, unmistakably detected on radar, yet to be peculiarly localized and preferential in their manifestation (that is, their appearance at times and places when and where they would be least likely to be detected, and their avoidance of level flight which would of necessity open them to observation by people along the way). The pattern in the "close encounter" cases is almost universal: a rapid descent to a landing or near landing, a stay of the order of only minutes, and the ascent, at usually a high angle, and disappearance either through distance or by some other means (it is often reported that at a height of a few hundred feet the bright luminosity vanishes) The choice of locale is statistically significant. The close encounter cases simply do not occur on the White

House lawn or between halves at the Rose Bowl game, but in desolate spots, generally some distance from habitation and where detection would be least expected. In a small percent of the close encounter cases, robot-like or human-like "creatures" are reported."[44]

Now, lets go back through the above quote sentence by sentence and, let's do it from the point of view of a hypothetical alien outpost on the Moon. The aliens are operating a remote sensing device as previously described. The following is a quick review of how it might work. Again I wish to emphasize that this is only my best guess extrapolating on present human technology. An alien technology hundreds, thousands, or millions of years in advance of our own would probably be beyond our comprehension.

A Quick Review

A particle beam with a strong positive charge and wave characteristics is encapsulated or ringed with electrons to the point that it is largely neutral and consequently is not deflected by Earth's or any magnetic field. The electrons exist in a quantum-entangled state with other electrons maintained in a synchrotron storage ring in the alien laboratory. The storage ring is housed in a special chamber that can utilize the synchrotron emissions of the ring to create a holographic image of what is being sensed at the beam's destination. The beam can be pointed and controlled quickly and with great accuracy. The total energy and velocity of the beam can also

---

[44] [15] p. xiii

be rapidly adjusted. The beam can reach Earth in a matter of seconds at some substantial fraction of light velocity and there, depending upon its energy, it penetrates Earth's atmosphere to a certain depth before gradual attenuation causes it to collide with air molecules creating a bright plasma display. With sufficient energy the beam can be made to reach all the way to Earth's surface. The entangled electrons orbiting the beam emit their synchrotron radiation and, in conjunction with the radiation of the plasma, "light up" the environment at the destination. This radiation bounces back to the beam and is received by the electrons. This active "sensing" of the environment is reflected in their entangled "brethren" back at the storage ring in the alien laboratory where alien technicians monitor a holographic image thus produced. The radiation at the target end is not entirely invisible and under certain atmospheric and ambient lighting conditions it (the virtual spaceship) can be easily seen by humans.

Now reading Hynek's statement again:

- Could it out-maneuver a jet fighter?   It could.

- Would it be silent?   Yes it would.

- Could it hover?   No problem.

- Could it attain high speed in seconds?   Easy.

- Would there be a sonic boom?   No.

- Could it change shape?   Quite possibly.

• Could it physically effect things?	Hot enough to burn!
• Would it show up on radar?	Yes.
• Could it descend and ascend at steep angles?	Along the beam.
• Could it disappear?	Turn off beam.
• Would luminosity diminish with elevation?	Less beam energy.
• Could human-like creatures be seen?	?

The aliens would probably be aware that their virtual craft is sometimes quite visible and, assuming that they wished to remain covert, they would avoid highly populated areas and prolonged visits for their "spaceship."

The modern era of UFO sightings began in 1947 when a pilot named Kenneth Arnold reported seeing nine "disc-like" objects flying in loose formation while on a trip between Chehalis and Yakima, Washington. He said they flew with an undulating motion like "a saucer skipping over water." This is believed to be the origin of the term "flying saucer." In addition, Arnold calculated the objects to be traveling at over 1700 miles per hour.[45]

If our aliens in their laboratory on the Moon moved their beam through a small angle it could cause the virtual spaceship to attain very high velocities in Earth's atmosphere. If the energy of the beam was kept constant during such a maneuver

---

[45] [15] p. 37

the spaceship would swing through an arc like a pendulum and appear to gain elevation at the end of the swing. This apparent gain in elevation would also be enhanced by the curvature of the Earth. If, however, the aliens wished to maintain a certain elevation relative to the Earth they would need to continually adjust the energy of the beam. This repeated adjustment in beam energy could cause the craft to fly as though skipping over water.

On the other hand, if the virtual spaceship was on some kind of automatic pilot and preset to fly at a given altitude relative to the Earth's surface, that could also explain an undulating motion as the craft passed over hill and dale.

Communication by Particle Beam

Now I would like to introduce a further speculation for the alien particle beam. I imagine a beam configured about the same as summarized above. The carrier beam is positively charged and has wave characteristics superimposed upon it, but now the wave is modulated to transmit information. This modulation could be of either amplitude or frequency much as we do for AM and FM radio waves. The information thus transmitted could be sequential and digital in either audio or text. An image could also be transmitted this way but it probably would not be holographic.

In his book *Where Is Everybody?- Fifty Solutions to Fermi's Paradox and the Problem of Extraterrestrial Life* [28], Stephen Webb lists solution #16 as, "They are calling but we do not know how to listen"[46] In other words we do not know how an alien civilization would choose to send such a signal.

---

[46] [28] p. 88

If they chose to send their signal in the manner suggested in the previous paragraph, we would not be aware of it at all. Such a signal would probably need a receiver of some kind and that receiver would have to be at a specifically targeted location. Maybe that is something the SETI people should chew on for a while.

I need to reiterate that I am probably in over my head with regard to even present human scientific knowledge. I am sure there must be some basic rules that would invalidate pieces and parts of the above speculations. But then, we would probably all be in over our heads when faced with an alien technology vastly older and more advanced than our own. However, I am most confident about the explanatory power of some sort of particle beam. Just how it would work is really an open question.

Chapter 12
Roswell

I gave serious consideration to completing this book without mentioning Roswell, the purported crash site of an alien spacecraft on June 14, 1947 in New Mexico. My proclivities run to science and engineering, not government cover-ups and conspiracy theories. In addition, I felt it was significant that Dr. Hynek did not mention this incident in either his first book [13] published in 1972 or his subsequent book [14] copyrighted in 1997.

But the legend surrounding the event has grown to such gigantic proportions in more modern times that it can no longer be completely ignored. This appears to be the result of the prodigious amount of research that has been focused on uncovering the truth of what actually happened. This research has been aided by the number of people who have since retired from the military or are approaching the end of their lives and no longer feel constrained about revealing what they know. Many of these people were only involved in some tangential way, but a few have direct experience. The specifics of the legend are now so well known that I don't feel compelled to detail them here.

Personally, I am inclined to accept the official Air Force version of what happened, but I admit that some big questions are still unanswered and as a result I am far from one hundred percent confident in that conclusion. The Air Force, in a detailed report,[47] lays out the results of their own internal investigation. They conclude there is no reason to suspect extraterrestrial involvement in the event. They admit the story released at the time of a downed weather balloon was, in fact, a cover-up. But they go on to say that what was being covered up was not the recovery of a crashed "flying saucer" as their initial news release actually stated, but a top secret endeavor called Project Mogul which was aimed at developing a balloon-carried, early warning system for a nuclear attack. The Air Force explanation was confirmed in an independent inquiry made by the General Accounting Office (GAO). A summary[48] of the GAO report includes this interesting detail:

"Bizarrely, some of the minor components of the balloon's payload had been supplied by a New York novelty company, whose fancy labels had been misinterpreted by eyewitnesses at the crash site as alien inscriptions."

Another reason I resist the scenario of a physical flying saucer crash is the basic theme I have been developing throughout the writing of this book. Even if there is an alien origin to some portion of UFO sightings, I maintain that the presence is more likely to be virtual — possibly holographic — and based on some kind of particle beam. A virtual space ship of the type envisioned would not crash land on Earth and this, by itself, could explain the lack of any physical evidence. However, it *is* possible that we are being visited by *physical* craft as well. When I say no physical evidence, a caveat is in

---

[47] MacAndrew, Capt. James. *The Roswell Report: Case Closed.* HQ United States Air Force. GPO, Washington D.C. 1997
[48] [03] p. 362

order: There is, at least, no physical evidence available for public viewing or for open scientific investigation.

The case made by many ufologists suffers from the usual pollution that is unfortunately rampant in this field. Key witnesses have, upon further investigation, been determined unreliable; classified government documents turned out to be forgeries; and there is outright perpetration of hoax. The waters were further muddied by the publication in 1997 of *The Day After Roswell*[01] by Col. Philip J. Corso (Ret.) The author, for reasons unknown, tries to insert himself directly into the middle of the Roswell legend. The book is replete with historical and factual errors and is possibly a rather transparent attempt to take commercial advantage of the considerable public interest that surrounds the event. In 1998 I wrote a review of Col. Corso's book as part of a memo to Gordon Chism and I have included a summary of the pertinent parts (Appendix C).

The whole subject of government cover-up is not an area in which I feel particularly qualified to even speculate. Other authors have pursued this line of investigation extensively. I recommend *UFOs and the National Security State* by Richard M. Dolan [08], an impressive work, and *Area 51-The Dreamland Chronicles* [06] by David Darlington, which is both informative and entertaining. If hard physical evidence *is* sequestered away in some secret location like Area 51, then this would, in my opinion, constitute a crime against humanity.

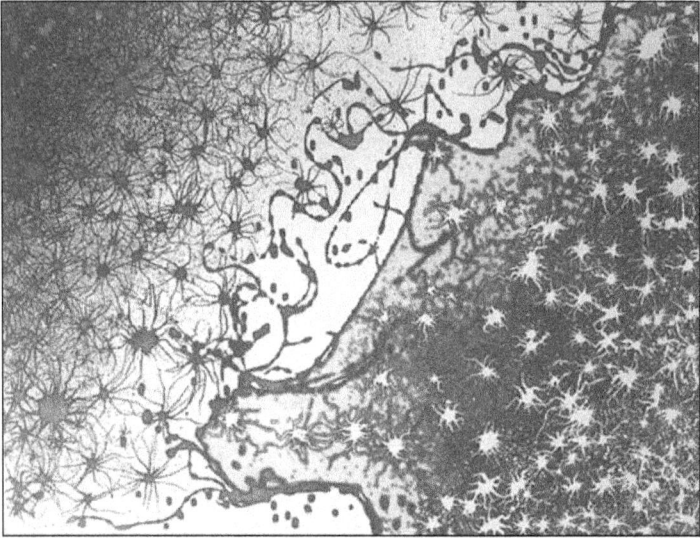

etching   Organic Life   R. L. Mason

Chapter 13
Back to the Sun

"Not only is the Universe stranger than we imagine, it is stranger than we can imagine."

Sir Arthur Eddington (1882-1944)
English astronomer

Life As We Know It

Much has been written about the requirements or pre-requisites for life. A planet must exist in the habitable zone of a star where water is predominately in a liquid state. The star must be in the habitable zone of a galaxy and must include a substantial inventory of the heavier elements, especially carbon. There should be a very long period (billions of years) of relative environmental stability where the elements are able to make innumerable random encounters while suspended in a solvent (liquid water in the case of Earth). Many other conditions are listed as desirable: A relatively large satellite is thought to be good for stabilizing the planet's rotation, and another giant planet in an outside orbit will help protect the nursery planet from bombardment by incoming comets and straying asteroids. And the planet should be geologically active, have a substantial atmosphere, and a magnetic field for protection from high energy particles, cosmic rays, and various types of undesirable radiation. Many of the above prerequisites are detailed in the book *Rare Earth: Why Complex Life is Uncommon in the Universe* [27] by Peter D. Ward and Donald Brownlee.

If all of the above sounds vaguely familiar, it should be no surprise for we have only one example to draw upon for our list of prerequisites.

Recently observations indicate that moons such as Jupiter's Europa and Saturn's Titan could serve as incubators of life. Europa is suspected of having a liquid water ocean under a layer of ice, and on Titan liquid methane may serve the role of solvent. In addition, it has long been speculated that silicon could play the same role that carbon does for life on Earth. These variations are considered possible because they are close analogs to the *single* circumstance we know is viable.

So much for the prerequisites, how do we recognize life when it actually happens? In other words, what is the definition of life? This is another subject on which much has been written. Here is a short, and not necessarily inclusive, list of often mentioned characteristics:

- Capacity for replication.

- Possession of a finite life span: It is born, lives, grows, and eventually dies.

- Responsiveness to its environment or outside stimuli.

- Capable of Darwinian evolution: being able to adapt and change over time or successive generations.

- Ability to move under its own power.

- Capable of metabolizing something in its environment as a reliable source of energy.

- Resistance to entropy: tending toward increased order rather than disorder.

Again, if this all sounds familiar there is a good reason, but what about life with a radically different origin?

Life As We Know It Not

I read with interest a synopsis[49] of the book: *Life Beyond Earth: The Intelligent Earthling's Guide to Life in the Universe*[09], by Robert Shapiro and Gerald Feinberg. The authors offer a much broader definition of life than is usual:

> "In seeking a broader view of the possibilities of life in the universe. . .we have started by framing a definition of life that is independent of the local characteristics of Earth life: *Life is the activity of a biosphere. A biosphere is a highly ordered system of matter and energy characterized by complex cycles that maintain or gradually increase the order of the system through an exchange of energy with its environment.* [emphasis added]
>
> An important feature in our definition is the identification of the biosphere as the unit of life. The history of life on Earth then becomes the tale of the continuous survival and evolution of the biosphere from its origin on the prebiotic Earth. Replication, and subdivision into organisms and species have been strategies adopted by our own biosphere to ensure its own survival but they need not be the methods used by an extraterrestrial biosphere."

---

[49] [29] pp. 168 to 171

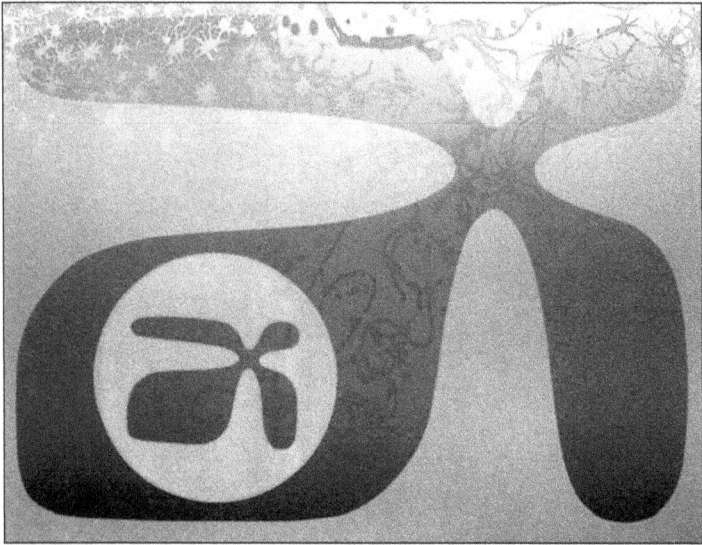

monoprint   Replication   R. L. Mason

This is what one might call "thinking outside the box," or maybe it should be called thinking inside the sphere. The authors then proceed to offer four examples of possible biospheres:

"1. Plasma life within stars. Such life would be based upon the reciprocal influence of patterns of magnetic force and the ordered motion of charged particles. It could exist within our own Sun. . ." [!]

2. Life in solid hydrogen. This could occur on a planet with a temperature of only a few Kelvin. Infrared energy would be absorbed and stored in the special arrangement of *ortho-* and *para-* hydrogen molecules.

3. Radiant life. Life would be based upon the ordered patterns of radiation emitted by isolated atoms and molecules in a dense interstellar cloud. Such clouds can have a long lifetime before they collapse. At a density of 104 atoms per cubic centimeter, they may last for millions of years.

4. Life in neutron stars. Such life would be based on the properties of the polymeric atoms whose existence was proposed by physicist Malcolm Ruderman (1974). It is possible that such polymer chains could store and transmit information in a way that bears an eerie similarity to the functions of nucleic acids."

As you can see from my interjection, I was immediately struck by item No. 1 above. The basic operating mechanism proposed comes quite close to a consistent theme running through this book, most particularly in Chapter 3. Items 2, 3, and 4 may be equally profound but I lack a sufficient frame of reference to make such a judgment. Thinking of the Sun as alive, maybe even aware and intelligent, is a concept so far out, so bizarre, that I almost laugh out loud every time I consider it. And yet. . . in a way, something about it is strangely appropriate.

Thought Experiment #3

In 1986 I decided to take a sabbatical of sorts and drive my VW poptop to Alaska on an extended camping trip. The trip started in late spring and as my latitude increased along with the declination of the Sun, the days grew longer and longer. Under such conditions one cannot avoid an increasing awareness of the Sun's influence on one's life. Upon reaching Alaska in mid-June, I arranged things so I would be at my furthest point north on June 21st or summer solstice. Consequently, on that date, I found myself at an elevated spot some sixty miles north of the Arctic Circle. Intellectually, I knew from my navy training in celestial navigation that the Sun would not dip below the horizon under such circumstances, but I wanted to see it for myself. I decided to watch the whole process for 24 hours, and, in addition, I dedicated that time to thinking about the Sun.

Thinking about a single subject for 24 hours is not as easy as it might sound. Occasionally I would drift off course, catch myself, and snap back to the subject at hand. From time to time I would fall asleep, then wake and attempt to start again where I left off. Initially, I attempted to recall every song lyric and line of poetry in my mental archives that mentioned the Sun in some regard. Then I reviewed all the steps and mathematics involved in calculating latitude from a Sun sight done with a sextant and known as LAN or Local Apparent Noon. Then I searched my memory for unique recollections involving the Sun. One of the earliest was me as a small boy frying ants on the sidewalk with a magnifying glass (I still remember the smell). Next, I remembered watching a sunrise or a sunset through binoculars from the bridge of a ship in mid-Pacific. The memory of what the world looked like during an eclipse of the Sun also came to mind. I wondered about the eleven year sunspot cycle and what could possibly cause such a thing. Focusing on religion, I ruminated about the Egyptian Pharaoh Akhenaton and his Queen Nefertiti who attempted to convert Egypt over to the worship of the Sun god *Ra*. I thought about Joshua in the Old Testament commanding the Sun to stand still and just what would have to happen for that to be a reality. I wondered if worship of the Sun might have been the original model for monotheism. Contemplating psychology, I remembered how, on a dull dreary day, your mood becomes suddenly elevated when the Sun breaks through the clouds and lights up the world. All this and much more passed through my mind during that 24 hour period.

At the end of all this observation and contemplation I came to the conclusion that the Sun is the single most important physical fact of our existence. We owe it *all* to the Sun, all life on Earth, all Earth's energy reserves, Earth itself. Anything that important is, at least, deserving of respect and maybe even reverence. So you can see how the seed cast by authors of item No. 1 above fell upon fertile ground.

Solar Life Speculations

The argument could be made that a close parallel exists between the life of a star and that of, say, a flowering plant. Both originate as a seed and grow by collecting materials or nutrients from their surrounding environment. Both "blossom" upon reaching a certain size and maturity. Both "blow" at the end of their "season" and scatter seeds or material that will become the genesis of the next generation. But is there Darwinian evolution in the case of the star? The plant, like all Earthlings, has a genome subject to occasional random mutation and its successive replictions are naturally selected for their ability to survive in a particular environmental niche. Stars *do* evolve because each successive generation includes a higher concentration of the heavier elements. These elements, formed by the fusion processes of the previous generation, then blown off into interstellar space and are contained in the dust clouds which produce the succeeding generations. Thus the first stars that formed after the big bang were made almost entirely of hydrogen with maybe a little helium and perhaps a smidge of lithium. Whereas the Sun, being possibly a third or forth generation star, while still primarily hydrogen, also includes an inventory of heavier elements. But this seems more akin to reincarnation than replication, and is there natural selection involved? What would a star be selected for? Natural selection is usually the result of competition for a finite quantity of resources. It's hard to see how that would apply to a star. But maybe it would apply to individual entities that make up the biosphere of a star. According to James Lovelock, founder of the Gaia theory, "Just as an animal is a cooperative assembly of living cells, so a planet can be considered as a living entity comprising a cooperative assembly of species." And, by extension, a star can be viewed this way also. The individual entities are plasma based rather than chemically based and they exist at much higher temperatures. This is somewhat like playing the same tune on a piano but several octaves higher.

My favorite science writer, David Darling, in his book *Life Everywhere: The Maverick Science of Astrobiology*[04], quotes Lynn Margulis, a biologist at the University of Massachusetts at Amherst and co-founder of the Gaia theory:

> "Life is a self-bounded system where the boundary is made by the material in the system. It's not a thing, it's a process, and these processes involve the production and maintenance of identity."[50]

If this is true and we apply it to a star such as the Sun, where we suspect some form of life, then we should expect to see boundary conditions, which define units with a cell-like character.

The first time I looked at the Sun through a telescope with the appropriate filter I was amazed to see it was not smooth as I had expected, but instead looked like the skin of an orange. The surface is mottled by what are known as granules where columns of plasma rise from the Sun's hot interior like bubbles in boiling water. What we see are the tops of convection cells that transport energy between the Sun's interior and its surface. Between these granules are bright points which are empty tubes of magnetic flux opening directly to the Sun's

---

[50] [04] p. 8 —originally from: Margulis, lynn and Sagan Dorian *What is life?* New York: Simon & Shuster (1996)

interior and are estimated to be as much as 125 miles long. As might be expected these strongly reminded me of linear accelerators.

In 2003, Romanian physicist Mircea Sanduloviciu and colleagues produced blobs of gaseous plasma that grew, replicated and communicated. They inserted electrodes into a chamber containing low temperature plasma of argon. An electrical discharge between the electrodes produced a concentration of ions and electrons to accumulate at the positively charged electrode. These concentrations spontaneously formed spheres which had a boundary of two layers with negatively charged electrons on the outside and positively charged ions on the inside. Inside each sphere was a nucleus of gas atoms. The amount of energy applied determined the size and lifespan of the spheres. Under the right conditions, they found the spheres could replicate by splitting into two, and could communicate by emitting electromagnetic radiation that made other spheres vibrate at a particular frequency. They referred to the spheres as gaseous *cells* and felt that there were important implications for the origin of life.[51]

Two things strike me as significant about this experiment: First, might this not be another way of neutralizing a positively charged beam of accelerated particles? Instead of an electrical discharge between two electrodes, what would happen if a stream of positively charged particles was passed through the chamber? Secondly, couldn't this be a model for the way a plasma life form is organized in a star?

Another phenomenon that may serve as an analog of a plasma cell is ball lightning and a short discussion of some

---

[51] http://www.newscientist.com/article.ns?id=dn4174

current thinking on this subject is included as Appendix D.

Should the Sun be a community of living — possibly intelligent — plasma cells, wouldn't it be fairly easy for them to organize a beam of the type previously proposed in this book for the purposes of remote sensing? Unlike us, they wouldn't have to build elaborate machines to originate such a system because they actually manifest the various principals involved and might do it as easily as we whistle a tune.

SETI scientists object to claims that the aliens they are trying to contact are already here. They cite the great the distances involved, the travel times required, and the insurmountable technical difficulties. The nearest star to the Sun is, after all, about 4.3 light years away. The *Sun*, however, is only 93 million miles away, and light from the "home star" takes only a little over eight minutes to reach the Earth. A beam of charged particles would take somewhat longer, but would still be quite short compared to interstellar travel times. So a UFO originating on the Sun as a particle beam would have a fairly easy time of it. Objections based on distance become moot.

The beauty of this speculation is its elimination of the two most prominent objections that have always plagued more standard UFO scenarios, the lack of hard physical evidence and the requirement of interstellar travel.

However, for extraterrestrial life based on the Sun, or which *is* the Sun, "alien" would seem to be the wrong word. This is, after all, the home star, and in a very real sense we are its progeny. It is our star . . . it is us. Could this account for the relatively benign quality of most reported UFO encounters?

In case the reader thinks I have gone off the deep end,

I still claim this is only speculation (and a wild one at that). A speculation only becomes a scientific hypothesis when it makes predictions that can be tested or verified (see Appendix A) and, frankly, I can't see any way that can be done in this case.

On the other hand, for a long time I have held a thought in the back of my mind that goes something like this: *If ever there comes a time when humans stumble upon or have revealed to them some great illuminating piece of knowledge so powerful and so compelling that all humanity immediately agrees to its obvious truth, then it will probably involve something that has been right in front of us all the time and which we are suddenly forced to see in an entirely new light.*

Chapter 14
Summary and Conclusions

In the preceding chapters I have speculated about UFO explanations in three main categories: natural causes, domestic technology, and alien technology. I personally, have not arrived at a point where I favor one of these three, and it is possible that all or some combination are involved to some degree. More time was spent on possible alien technology because, as stated in the Introduction, that is the *key* question, and it is also the most fun to think about. However, a consistent theme runs through this book, cuts across all three categories, and I believe has been largely overlooked by other authors writing on this subject. In my opinion the involvement of some kind particle beam can explain many of the commonly mentioned characteristics listed in **Table II** *and the dearth of physical evidence.*

Discovering the true nature of UFOs is not an easy task for science even if scientists *were* inclined to tackle it. If the *natural causes* speculation is true, it represents a particularly tough observational problem. Where do you set up your equipment? The phenomena appears to be random as to location, and may depend on the coming together of several events or conditions all of which have a very low probability individually. If the *domestic technology* speculation is true, UFOs may be the result of highly classified developments in which every effort is being made to keep them away from the public eye. The United States alone has a huge "black budget." Who knows what's going on in there? If the *alien technology* speculation is true, then we have a *really* tough problem. How do you make observations of entities who may wish to observe but, apparently, do not wish to be observed, especially if their technology is vastly superior to our own? And finally there is the possibility that combines the *natural causes* speculation with the *alien technology* speculation. A possible life form in our presence so different from us that we don't even recognize it as life. How does science approach that kind of problem? One way to start might be to ask ourselves what is it we experience on a fairly regular basis that we can't explain, something for which we are almost in denial? UFOs seem to fit that description.

Observations *are* being made fairly frequently. They are not "scientific" observations; they are haphazard, anecdotal, and many are unreliable. However, even if only a very small portion are valid their sheer number is enough to convince me (as it was for Dr. Hynek) that *some physical phenomena is being observed.* The scientific establishment, on the other hand, seems to have decided to ignore the entire question hoping that maybe it will just go away. So here we have the ironic spectacle of *empirical* science spending an inordinate amount of time on things like superstring theory and M-brane theory which offers little or no hope of ever making any kind of verifying observation, while the phenomena in question is loaded with observations and receives almost no scientific attention at all. Something is not quite right about that. As I have pointed out in Appendix A, the cognitive cycle, which is the heart and soul of science, begins and ends with observation.

With some significant exceptions, science has come very close to abdication on the subject of UFOs, but that leaves the door open to laymen like me who are, in general, scientifically literate, but unaffiliated and far from expert in any particular field. Not locked within the structure of academia or some national laboratory, and free from the intimidation of peer review, one has a certain amount of "maneuverability" even if lacking *gravitas*. Any way you cut it, the question, "Are we alone in the universe?" must be one of the premier intellectual questions of our time. If this book contributes in some way, however small, toward answering that question, I will die a happy man. Thank you for reading.

Epilogue

After finishing this writing project, I set aside what I hoped was my final version of the text with the intent of letting it "cool" for a period. When a sufficient time had lapsed, I picked it up and tried to read it again with fresh eyes. Generally speaking, I was satisfied with its development, but one additional thought occurred to me: *Should there come a time in the near future when we are actually confronted with the presence of a physical spacecraft from another star system, what are we likely to discover?*

Given our own rather remarkable progress during the last ten to twenty years in artificial intelligence, nanotechnology, robotics, etc., one can only guess at what the next hundred, thousand, or million years might bring. One speculation worth exploring is the possibility that we may give up our biological existence altogether. Artificial life may eventually take over where biological life ends, and this may be an inevitable outcome of any highly evolved technological society. If true, then we may find that an alien spacecraft arriving from another star system may not contain autonomous beings at all — an unnecessary complication. An alien intelligence could build itself into any configuration to serve whatever task is at hand. Thus, the alien may be built right into a spacecraft. The spacecraft *is* the alien and vice versa.

Such a development would certainly eliminate some of the difficulties of interstellar space flight of which we are just now becoming aware. It would only be necessary to "turn off" the intelligence or place it in a "safe mode" for voyages of tens, hundreds, or thousands of years. No need to worry about boredom, for instance.

And that is truly the last speculation of this book.

Appendixes

Appendix A
The Philosophical Roots of Science

The word "science" comes from the Latin *scientia* "knowledge," but in its modern English usage it has come to mean more than that. Also implied is a system or method based on observation. Here is my definition:

> *Science is an empirically based system for the acquisition, compilation and dissemination of knowledge about the physical Universe.*

Inclusion of the word "empirical" denotes observations acquired by the senses. The branch of philosophy dealing with this is known as epistemology. What follows is a description of a particular epistemology, the one that makes the most sense to me.

First, a basic assumption: all objective knowledge, or any thought that carries intelligence, can be put in the form of statements. It can be communicated, or lifted from one individual's brain, put into words and transmitted to another's brain orally, or in writing. And this is possible regardless of truth or falsehood, how you got it, or where it came from.

With all objective knowledge or intelligence in the form of statements, we can examine their structural content. We can do some sorting and pigeonholing. The first distinction to be made is between analytic and synthetic statements:

An analytic statement is knowable (either true or false) without reference to the world. You know it in your head (*a priori)* and no observation is necessary.

Example: All brothers are male siblings.

We know this statement is true because the definition of brothers *is* male siblings. It is true by definition. You could take the words "male siblings" out of the statement and plug in "brothers" and the statement would read: all brothers are brothers. We can tell this is true simply from the structure of the statement itself. There is no need to go out and look at brothers. It doesn't tell you anything about the nature of brothers. Another name for such a statement is a tautology. Some tautologies are quite famous.

Example: What will be will be.

At first, this sounds like a profound statement, as if it were telling you some basic truth about the nature of reality. In fact, it *is* true, *absolutely* true, but its truth comes from the structure of the statement not from some long, hard-earned experience with the world.

The truth (or falsehood) of tautologies and all other analytic statements is necessarily absolute; they are set up by us to be so. They are true by definition but they do not address the nature of reality.

Synthetic statements, on the other hand, *do* speak of the nature of the world outside your head.

Example: It is raining.

We don't automatically know *a priori* the truth or falsehood of that statement. Its' truth or falsehood is contingent upon whether, in fact, it *is* raining and that determination requires a verification process. we must go to the window and look out, or we must listen for the sound of rain on the roof, or we must feel the dampness in the air. In other words, we must gather sense data from the world in order to confirm or deny the statement and only then (*posterior*) can we determine its truth or falsehood. Moreover, such a determination is not absolute as it is with analytic statements. Sense data is fallible and not every observation necessary for a determination can always be made. If it is night, going to the window and looking out may not help. The sound we thought was rain on the roof might, in fact, just be some dry leaves blowing about, and the dampness in the air may come from the kettle on the stove.

Consequently, the determination of the truth or falsehood of a synthetic statement must always be expressed as a probability. The truth of a synthetic statement may be very highly probable.

Example: Gravity exists.

But since every possible observation has not been made (and never will be) the existence of gravity must remain very highly probable, but *not* absolute.

On the other hand, even though the truth or falsehood of synthetic statements cannot be determined absolutely, they *do* tell you something about the world. They are useful in dealing with reality. Exhibit A lists the characteristics of analytic and synthetic statements and displays them in summary form.

Since it is very highly probable we human beings exist, and since it is also very highly probable the world, nature, and the universe also are a reality, it is important that we develop a system for determining the relative truth (or falsehood) of

EXHIBIT A: Characteristics of Analytic and Synthetic Statements

BASIC ASSUMPTION: All objective knowledge can be put in the form of a statement

The two types of statements	The way the statement is knowable	How a conclusion is reached about truth or falsehood	The degree of certainty	Comments
ANALYTIC *Example:* "All bachelors are unmarried"	**A PRIORI** • self explanatory • without reference to the world • in your head	**NECESSARILY** • by definition • logic	**ABSOLUTE**	• the rules of logic tell us nothing about the nature of the world • they are relationships which we define to be true in our minds • they are the tools, not the product
SYNTHETIC *Example:* "All metals expand when heated"	**POSTERIORI** • by observation • through reference to the world outside your head	**CONTINGENTLY** • depends on what is observed	**PROBABILISTIC** • absolute synthetic knowledge is not possible because: 1) we are limited in space & time 2) our senses are imperfect 3) we each bring our own set of individual references	• all knowledge of our existence and the universe outside our head is in this form

122

synthetic statements, a kind of test of their reliability. In fact, we have done just that and we couldn't have succeeded in nature to the degree we have if such a process had not become a manifest part of the human experience.

How do we arrive at important synthetic statements in the first place? There is a process labeled induction by enumeration which begins with observations of reality. Initially, this is a random process, but eventually observations start to lump themselves together into categories and frames of reference.

For example, I see an animal. Eventually I see another that looks like the first, and then I see a third and a fourth. These observations become a frame of reference centered on that kind of animal. I give it a label, "cat." I notice the first four cats all had tails. I see a fifth and a sixth cat. They have tails too. And now I make an inference. Based on my specific observations, I make a generalized statement about cats:

Example: All cats have tails.

This kind of synthetic statement is called a hypothesis, and the process is induction by enumeration or inductive logic. Inductive logic always moves from the specific to the general and is synthetic in nature. Having made the hypothesis, I now treat it as true. But if I am realistic, I realize its truth is only probable to a degree, and that degree is tied directly to the number of observations I have made (namely six). However, I proceed merrily along as if the hypothesis were true. I make a prediction. I conclude that the next cat I see will have a tail. This is deductive logic. Deductive logic always moves from the general to the specific. I see a seventh cat. It has a tail. This observation confirms my hypothesis and my confidence level rises. I continue on in this manner growing more and more confident until — oh no!— a cat *without* a tail! Woe is me! But all is not lost. I really do not need to start all over at the beginning. My observations are still good; it's just my hypothesis that is flawed. It needs a little work. How about this:

Example: *Most* cats have tails.

This process of moving from the specific to the general, from the general to the specific and back around again is the cognitive cycle and is the basis for what is known as the Scientific Method. Scientists call hypotheses that are very highly probable *facts*. But even facts are not held to be absolute. A diagram of the process is shown in Exhibit B.

We think this way naturally, without making a conscious effort. The Scientific Method is simply an elaborate formalization of the way our heads work automatically. I would go as far as to say *any intelligence that gathers knowledge of the Universe empirically (through the use of senses) would necessarily function in the same way.*

Exhibit B is a fairly good representation of the cognitive cycle, with one shortcoming. It shows the realm inside your head (the brain) as isolated and separate from the universe outside. When, in fact, you and your brain are *in* the Universe as well. Thus we are able to make observations of our own mental activity, which may be at the root of consciousness.

The examples I have used are absurdly simple ones, and I don't mean to give the impression that the process is always so straightforward. It can become immensely complex, but no matter how many variations on the theme you find, it's still the same basic process.

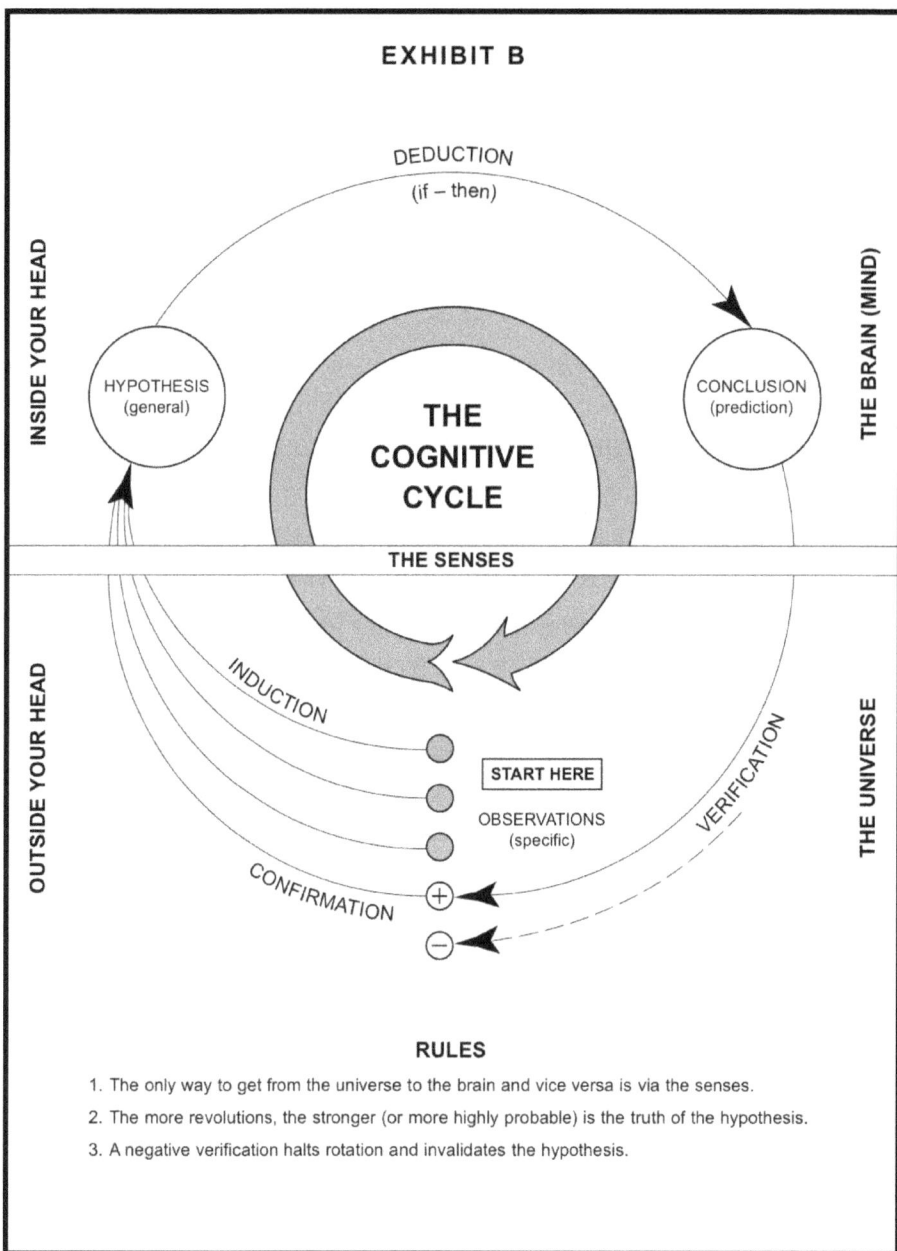

EXHIBIT B

DEDUCTION
(if – then)

INSIDE YOUR HEAD

THE BRAIN (MIND)

HYPOTHESIS
(general)

THE COGNITIVE CYCLE

CONCLUSION
(prediction)

THE SENSES

OUTSIDE YOUR HEAD

THE UNIVERSE

INDUCTION

VERIFICATION

START HERE

OBSERVATIONS
(specific)

CONFIRMATION

$\oplus$

$\ominus$

RULES

1. The only way to get from the universe to the brain and vice versa is via the senses.

2. The more revolutions, the stronger (or more highly probable) is the truth of the hypothesis.

3. A negative verification halts rotation and invalidates the hypothesis.

Up to this point I haven't said anything very controversial. I think most philosophers would largely agree with what I've stated so far. But there is a big split in philosophy and we are rapidly approaching that point. So, if you will look again at Exhibit A, you will note I have drawn a heavy line between the characteristics of analytic statements and the characteristics of synthetic statements. Philosophers generally lumped together as empiricists say there is no crossing that line. Synthetic statements cannot be knowable *a priori*, (in your head without reference to the world). Philosophers usually gathered under the general heading of rationalists claim at least one such statement and they may claim more. They go on to claim that starting with such a statement as a given and using pure reason (deductive logic), all of reality can be derived.

Such statements are usually termed metaphysical (beyond physical) or supernatural which conveniently releases them from the constraints Mother Nature places on the rest of us. Probably the classic metaphysical statement of all time is: "God exists." So much controversy has swirled around just this one metaphysical statement that the various positions have their own labels. There are theists, atheists and agnostics.

A theist is a rationalist who claims *a priori* knowledge that the statement "God exists" is true. An atheist is *also* a rationalist, but claims knowledge that the statement is false. All true empiricists are agnostic. They claim knowledge is not possible concerning the truth or falsehood of this statement, or any metaphysical statement, since such statements do not lend themselves to observation and verification by the senses. An agnostic would maintain that such statements are matters of faith, not knowledge.

Agnosticism is really a much broader category than most people realize. For instance, one can have *faith* that "God exists" is true and still be an agnostic. Or you can *believe* the statement is false and be an agnostic. Only the claim of *knowledge* will move you out of the agnostic category one way or the other.

Other metaphysical statements abound and usually concern the existence of things such as souls, spirits, gods of all kinds, large and small, ghosts, poltergeist, goblins, etc. In fact, the field is wide open. You can dream up your own. All you have to do is claim knowledge that they exist, but that they do not lend themselves to sense observation. Talk about free enterprise! This is *really* an unregulated industry!

The trouble with human beings is we have very active imaginations and, although this can be entertaining and useful when properly channeled, it often runs wild and tempts us to create answers where voids in our knowledge exist. These answers form the dogmas that differ from one religion to the next. They differ because they are not based on common experience of the senses. Whole wars have been fought and unspeakable atrocities committed in the name of these differences. As a species, and as individuals, we need to discipline ourselves to reserve judgment where voids in our knowledge exist. We need to learn to say, "I don't know" without feeling inadequate. We must be patient.

In conclusion: all objective truths or "facts" are simply matters of high probability and all knowledge about the physical nature of the Universe is of this nature.

"I think that in discussions of physical problems we ought to begin not from the authority of scriptural passages, but from sense-experiences and necessary demonstrations…"

Galileo Galilei

Appendix B
E-mail to Sandia National Laboratories

To: XXXX XXXXXXXX, Sandia National Laboratories
From: Rob Mason, Mendocino, CA
Subject: Sandia Fireball Network

Greetings, I am involved in a writing project that causes me to want more information about the subject network. What I know, at present, comes mainly from an article I read in a May 2004 article of ASTRONOMY magazine, entitled *All-sky fireball network.* The writing project is an attempt to update, and is loosely based on, the book authored by the late astronomer J. Allen Hynek entitled *The UFO Experience: A Scientific Inquiry.* Published in 1972, this book is generally acknowledged to be the most influential work ever published on the subject of UFOs. I know it influenced me because until I read it I was a total skeptic regarding UFOs, (at least those with a possible alien origin) and afterward I became . . . well. . . agnostic and very curious. My personal background is that of an engineer with a broad working experience including some exposure to particle accelerators.

Dr. Hynek's book is influential because of his scientific approach, but unfortunately almost all the data he had to work with was anecdotal in nature. In addition, a lot has happened since 1972. The new book is tentatively entitled *The UFO Experience Reconsidered: Speculations, Scientific and Otherwise.* Speculations are offered in three main areas;

natural causes, domestic technology, and alien technology. It is my desire to include a chapter on suggested ways to get beyond anecdotal evidence, and that is why the above-mentioned article caught my eye.

So, In light of the above, my three main questions are:

1. Does the Sandia Fireball Network ever record objects (fireballs?) that;
   a.) change course during the observation, or
   b.) move too slowly to be a meteor, or
   c.) stop and hover, or
   d.) accelerate instead of decelerate, or
   e.) exhibit any of the other characteristics commonly mentioned in association with UFOs?

2. If so, is this information collected and tabulated in any way?

3. If so, is it available to the general public?

Thank you so much for your time.

Appendix C
The Day After Roswell - A Book Report

MEMORANDUM

Date: December 15, 1998
To: Gordon Chism
From: Rob Mason
Subject: *The Day After Roswell* — a book report

First, let me say, I agree with your assessment of Col. Corso. He is, indeed, a "limited" person. His need to continually puff himself up is quite apparent. . .

. . . It just so happens that my active duty tour [U.S. Navy] coincides exactly with Col. Corso's time in Army R&D at the Pentagon. This does not give me any special insight into the veracity of Col. Corso's account except possibly a general feel for the climate within the military during that period.

To begin, let us assume, for the sake of argument, that the Colonel's story is in fact true; that things happened just as he said they did and the whole thing is a huge cover-up. The military, in their wisdom, classified everything concerning the affair "above TOP SECRET" —a decision for which the Colonel expresses his approval on several occasions. If this is so, it must *still* be highly classified, because if it had ever been declassified we surely would have heard about it. The Colonel, on the other hand, appears to have changed his mind because didn't he just blab the whole thing —in writing?

When I was in the Navy, I was cleared for TOP SECRET (cryptographic), and at one point I was TOP SECRET Material

Control Officer for the ship I was on. Again, this does not give me any special insight into the Roswell affair, but I do know that if you reveal classified material, even after you have left the service, you can be prosecuted. Is Colonel Corso being prosecuted? I haven't heard anything, if not, why not? Two possible explanations occur to me:

1) The Colonel is not being prosecuted, because to do so, the military would have to admit too much. By not prosecuting they are counting on the Colonel being generally considered to be some kind of nut, and they hope the whole thing will just blow over; or

2) The Colonel is not being prosecuted because the whole thing is a fabrication and no secrets were revealed in the first place.

Another point that bothers me is the descriptions of the aliens themselves. As described, the aliens would fall dead center into the whole science fiction/Hollywood mythology about "little green men." Well maybe they weren't green — but close — they were gray. This is a little hard for me to swallow. The probability that a sentient species from a distant star system, with a whole separate evolution, would in any way resemble us has got to be vanishingly small.

It is said of the ancient Greeks that "they turned their gods into men and their men into gods." And I am mindful of the fact that Michelangelo, when representing God on the ceiling of the Sistine Chapel, painted a human being. Science has forced a long series of demotions on us since those days, but are we still falling into that trap?

IF there is an intelligent, technologically advanced species out there somewhere, THEN they must look something like us.

In spite of what you see on *Star Trek*, I — don't — think — so. Okay, this *is* a problem. Here are two possible explanations:

1) The aliens look somewhat like us because they *are* us [from the future]. Corso himself hints at this on a couple of occasions, but seems to reject it in the end. This will require a complete overhaul of the laws of physics, but . . .maybe so; or

2) The whole thing is a fabrication and Corso has simply tapped into the popular myth in order to increase his credibility with the masses.

Another bothersome point involves the intentions of the aliens. The Colonel, being a military type, needs to have enemies so he automatically assumes that the aliens are hostile. The "hostile" acts he lists are the mutilations of cattle, violating our air space, disrupting our communications and, oh yes, abductions! Well, hmm. If the aliens are hostile then they must have weapons, and if they have weapons, then we would have to assume that their weapons are as superior to ours as their spacecraft is to our aircraft. So why isn't there some instance of a direct attack? It would seem that they should be capable of overwhelming force, but instead we get this other stuff. Here, again, are two possible explanations:

1) There is no direct overwhelming attack because the aliens really *aren't* hostile. They are just

curious. They are on a scientific mission and we are part of their study; or

2) There is no direct overwhelming attack because that is something that everyone would be aware of immediately. Corso can only maintain his little fiction by keeping things on a subtler level that only the cognoscenti, civilian and military, know about.

Have you ever heard of the scientific maxim called Occam's Razor? It reads [paraphrased]:

"When given a choice between several explanations for a given phenomenon the one which is the simplest and most straight forward is the one most likely to be true."

Certainly, accepting as true that we are being visited by aliens from another star system (or from the future) whose spacecraft defy the laws of physics as we know them, and whose very existence has been the subject of a massive government cover-up which has been maintained for over fifty years, is very complex. The alternative is that Col. Philip J. Corso (Ret.) is a liar and that's a fairly simple and straightforward explanation.

Of course, if you are a career officer in Army intelligence you don't tell lies, you disseminate disinformation. "Disinformation" is a word that Corso uses on numerous occasions in this book. It is one of the standard tools of intelligence organizations and I assume that Corso is practiced in its use.

Much of the latter portion of this book is a rundown of various projects that were being pursued by Army R&D while Corso was at the Foreign Technology Desk in the Pentagon. After reading through a number of these I got the feeling that these were summaries he had written at some much earlier time, which he had pulled out of his archives and modified for inclusion in this book. Often it seemed to me that the only change he made was to insert the words "and the extraterrestrials" everywhere that the original text mentioned the USSR. Sometimes this phrase was left dangling at the end of a sentence or paragraph like an afterthought.

There is one respect in which Corso is very fortunate. He has lived to a ripe old age (over eighty). He mentions a lot of real people by name in his book, but most of them are now deceased. One exception is Senator Strom Thurmond who is still in the Senate at a very ripe old age (96). It is my understanding that Senator Thurmond originally wrote a forward to Corso's work, but later thought better of it and withdrew it.

Above, I mentioned the possibility that Col. Corso is a liar, plain and simple. Perhaps that is a bit harsh. I'll give him the benefit of the doubt and say that his book, *The Day After Roswell* impresses me as the deluded ramblings of an old man.

Appendix D
Ball lightning

Ball lightning has been a puzzle for many years. Its very existence was questioned for a long time. Recently, however, John Abrahamson, Associate Professor of Chemical Engineering at the University of Canterbury in Christchurch, New Zealand, claims to have come up with an explanation that sounds reasonable. He points out that when lightning strikes soil, it often forms a "fulgurite" which is a roughly cylindrical fused column of material that may be dug up and examined after the fact. Immediately after the strike, the column is molten, but it cools from the outside where it is in contact with the surrounding soil and, in the process of cooling, can form a tube. Soil materials in a gaseous state may be expelled out of the tube and back into the atmosphere. If the materials in the soil include silicon oxide and carbon, metallic silicon vapor may result and condense into long strings of silicon nanospheres which, in turn form a gaseous sphere in the atmosphere. Professor Abrahamson explains "You have quite a robust structure, which continues to oxidize, and stays hot and visible." Oxide on the surface of nanospheres slows the process down until each particle finally runs out of metal.

It seems probable to me that the materials in the soil, which are expelled from the fulgurites in a gaseous state, have been ionized by the energy of the lightning strike and this promotes the formation of the nanospheres as well as the "ball" in the atmosphere. Thus as the materials return to equilibrium via oxidation, electromagnetic radiation is given off in the form of light, and consequently what we have is a form of plasma cell.

[Bibliography]

[01]Corso, Philip J. with William J. Birnes *The Day After Roswell.* New York: Simon & Shuster, 1997

[02]Bracewell, Ronald N. *The Galactic Club: Intelligent Life in Outer Space.* New York-London: W.W. Norton & Co., 1974

[03]Darling, David *The Extraterrestrial Encyclopedia: An Alphabetical Reference to All Life in the Universe.* New York: Three Rivers Press, 2000

[04]Darling, David *Life Everywhere: The Maverick Science of Astrobiology.* New York: Basic Books, 2001

[05]Darling, David *Teleportation: The Impossible Leap.* Hoboken, New Jersey: John Wiley & Sons Inc, 2005.

[06]Darlington, David *Area 51: The Dreamland Chronicles.* New York: Henry Holt and Co., 1997

[07]Desmond, Leslie and Adamski, George *Flying Saucers Have Landed.* New York: The British Book Centre, London: Werner Laurie, 1953.

[08]Dolan, Richard M. *UFOs and the National Security State*. Charlottesville: Hampton Roads Publishing Company, Inc. 2002.

[09]Feinberg, Gerald & Shapiro, Robert *Life Beyond Earth: The Intelligent Earthling's Guide to Life in the Universe.* New York: William Morrow and Co. Inc., 1980

[10]Good, I.J. (ed) *The Scientist Speculates: An Anthology of Partly-Baked Ideas*. New York: Basic Books, 1962.

[11]Gould, Stephen Jay *Wonderful Life: The Burgess Shale and the Nature of History*. New York-London: W. W. Norton & Company, 1990.

[12]Green, Brian *The Elegant Universe*. New York-London: W. W. Norton & Co. 1999.

[13]Hynek, J. Allen *The UFO Experience: A Scientific Inquiry*. New York: Ballantine Books, 1972.

[14]Hynek, J. Allen *The Hynek UFO Report*. New York: Barnes & Noble Books, 1997.

[15]Jacobs, David Michael *The UFO Controversy in America*. Bloomington: Indiana University Press, 1975.

[16]Krauss, Lawrence M. *The Physics of Star Trek*. New York: Harper Collins Publishers, 1995

[17]Levin, Janna *How the Universe Got Its Spots*. Princeton: Princeton Press 2002.

[18]North, Gerald *Observing the Moon*. Cambridge: Cambridge University Press, 2000.

[19]Ruppelt, Edward J. *The report on Unidentified Flying Objects*. Nashville: Source Books Inc., 2002.

[20]Russell, Bertrand *Wisdom of the West*. New York: Cresent Books Inc., 1959

[21]Sagan, Carl & Agel, Jerome *Cosmic Connection*. New York: Doubleday, 1973

[22]Sagan, Carl & Page, Thornton (eds.) *UFO's: A Scientific Debate*. New York-London: W.W. Norton & Co., 1972

[23]Sagan, Carl *The Varieties of Scientific Experience.* New York: The Penguin Press, 2006

[24]Smolin, Lee *The Trouble with Physics.* Boston-New York: Houghton Mifflin Co. 2006.

[25]Sturrock, Peter A. *The UFO Enigma: A New Review of the Physical Evidence.* New York: Warner Books, 1999.

[26]Ward, Peter D. *Life As We Do Not Know It.* New York: Penguin Books, 2005.

[27]Ward, Peter D. & Brownlee, Donald *Rare Earth: Why Complex Life Is Uncommon in the Universe.* New York: Copernicus Books, 2000.

[28]Webb, Stephen *Where is Everybody?- Fifty solutions to the Fermi Paradox and the Problem of Extraterrestrial Life.* New York: Copernicus Books, 2002.

[29]Zuckerman, Ben, and Hart, Michael H. *Extraterrestrials: Where are They?* Cambridge: Cambridge University Press, 1995.

Index

daylight discs (DD), 6, 22
decoherence, 49–50
deductive logic, 123–126
dimension, spatial, 57–61
disc UFOs, 9, 18
Discover (magazine), 21
"disinformation," 134
Dolan, Richard, 97
domestic speculations, 26–31,
    72–73, 111–112
Drake, Frank, 34–37
Drake's Equation, 35–37
Durie, David, 67, 69
dust clouds, 62

Earth
    aliens navigating in relation
        to, 93
    characteristics of life on,
        101
    dependence on the Sun,
        63–64
    diffusion of solar flares,
        20–22
    as future home to aliens,
        79–81
    as galactic entertainment
        park, 82
    magnetic field diffusion,
        20–22
    as reached by particle
        beams, 91
Einstein, Albert, 35, 38, 60
electrical equipment, and solar
    activity, 22
electromagnetic radiation, 41,
    42, 62, 108, 138
Elegant Universe, The
    (Greene), 59
elliptical UFOs, 9, 18

empiricists/empirical science,
    112, 119, 126
engines, 10, 22
entanglement. *See* quantum
    entanglement
epistemology, 119
European Space Agency's
    SMART-1 mission (2005-
    2006), 51
evidence gathering, 85–88
evolution, 54, 100, 106

Falon Naval Air Station, 14
Farmer, Mark, 30
Feinberg, Gerald, 101, 103
Fermi, Enrico, 77
Fermi's Paradox, 77, 78
Feynman, Richard, 57
fireball networks, 85–86,
    129–130
fluorescent qualities, 9, 22, 43
Flying Saucers Have Landed
    (Adamski), 2
fourth dimension (4D), 58
fraudulent claims, 1–2
fulgurite, 137

Gaia theory, 106, 107
galaxies, 37, 62, 79–80
Galileo Galilei, 128
General Accounting Office
    (GAO), 96
Gill, William Booth, 67–75
glowing quality, 9, 12, 13, 18,
    22
Gould, Stephen Jay, 54
government cover-up, 95, 97
gravitational accretion, 62
Green Bank Conference, 35
Greene, Brian, 59

Groom Lake (NV), 26, 30
Gulliver's Travels (Swift), 63

habitable zones, 79–80, 99
Hatch, Larry, 23
"Here Comes the Sun" (article;
    Mackensie), 20, 21
holographic images, 43, 73,
    82, 90, 91
"hostile" aliens, 133–134
hovering prototypes, 9, 89
*How the Universe Got Its
    Spots* (Levin), 61
humanoids
    as characteristic of close
        encounters, 6, 53–55, 90
    in Corso's Roswell account,
        132–133
    as designer beings, 56, 74
    as futuristic humans, 55–56,
        133
    as myth, 55, 133
    in New Guinea sighting, 70,
        72, 74
hydrogen, 103
hydrothermal vents, 65
Hynek, J. Allen, 2, 5–8, 11, 17,
    23, 33–35, 38, 53–55, 71,
    72, 85, 89–91, 95, 112, 129
hypotheses, 123–124

infrared lens, 87, 88
intragalactic treaty theory, 79,
    80
interstellar space travel, 109,
    115–116
    See also teleportation

Jones, Ed, 25
Jones, Robert T., 25–28

Jupiter, 64, 81
Jupiter's Europa, 100

Kubrick, Stanley, 50

Levin, Jana, 61
*Life Beyond Earth: The
    Intelligent Earthling's
    Guide to Life in the
    Universe* (Shapiro and
    Feinberg), 101
*Life Everywhere: The
    Maverick Science of
    Astrobiology* (Darling), 107
life forms, prerequisites for,
    99–101, 103–104
Life Magazine (periodical), 14,
    15, 17
"Life on Stellar Surfaces"
    (essay; Shapley), 63
lightning, 137
Lilliputs, 62–65
linear particle accelerators
    (linac), 19, 28, 29, 31
locale, 89–90, 92
Loop Quantum Gravity theory,
    60
Lovelock, James, 106–107
luminous qualities, 9, 12, 13,
    18, 22
lunar outposts, 49–52, 81, 92

M-brane theory, 59–61, 112
Mackensie, Dana, 20
macroverse, 61–62
magnetic field, Earth's, 20–22
Mahood, Tom, 29–31, 43
Margulis, Lynn, 107
Mars, 81
metaphysics, 59–60, 126–127

www.ingramcontent.com/pod-product-compliance
Lightning Source LLC
Chambersburg PA
CBHW031938190326
41519CB00007B/579